# NORTH STAR TO SOUTHERN CROSS

# NORTH STAR TO SOUTHERN CROSS

Will Kyselka &
Ray Lanterman

THE UNIVERSITY PRESS OF HAWAII
Honolulu

Manufactured in the United States of America

Designed by Ray Lanterman

**Library of Congress Cataloging in Publication Data**

Kyselka, Will.
North Star to Southern Cross.

Bibliography: p.
Includes index.
1. Stars. I. Lanterman, Ray. II. Title.
QB801.6.K95 523 75-37655
ISBN 0-8248-0411-2
ISBN 0-8248-0419-8 pbk.

# CONTENTS

# PREFACE

North Star to Southern Cross came out of my ten years' experience in working with audiences at the Bernice P. Bishop Museum Planetarium in Honolulu, and with thousands of young people at school camps by the sea.

My purpose in writing the book has been to provide a framework for understanding the heavens. It is intended for a broad range of readers: The student just beginning to discover the excitement of the starry sky, and for the serious amateur as well, who wants a ready guide to the stars of each month and the brighter Messier objects. It is for the city dweller for whom technology may have dimmed the sky, but not his curiosity, and who wonders, "What's that bright star?" It is for the hiker in mountain or desert, and for the boatsman far out to sea who finds a sky so filled with stars that he is momentarily lost in its brilliance.

Many persons have contributed to the development of this work, some through personal conversations, and others in their writings. Quotations in the third chapter are from R.H. Allen's classic work, *Star Names.*

My thanks to Walter Steiger, chairman of the Department of Physics and Astronomy at the University of Hawaii, for introducing me to the intricacies of planetarium work. And to Bishop Museum astronomer George W. Bunton, from whom I have learned how effective human imagination can be, when coupled to a planetarium instrument, in making understandable the ideas of astronomy. His commitment to presenting science in a clear, precise way has, over the past three decades, stirred the imaginations of planetarium audiences in Los Angeles, San Francisco, and Honolulu to the splendor—seen and unseen—in the sky.

An important astronomical center is developing at the top of an old volcano in the middle of the Pacific. Quiescent since the end of the last Ice Age, majestic Mauna Kea rises 2½ miles above the sea and into the thin air where the sky is clear, dark, and cold, for the world's finest unobstructed infrared view of the universe.

Several astronomers who use the Mauna Kea Observatory's 88-inch telescope have had a part in shaping this book. John Jefferies, director of the University of Hawaii's Institute for Astronomy, has encouraged our project; and, coming from "down under," he is pleased that we've included the sky all the way to the south celestial pole. Solar astrophysicist Frank Orrall plotted the precessional circles for both the northern and southern hemispheres; he also read and commented on parts of the manuscript. Franklin Roach, an astronomer interested in sky glow, and also a planetarium lecturer, has helped with several diagrams and has given encouragement at a crucial time. Dale Cruikshank, associate director for the Mauna Kea Observatory, and astronomers Ann Boesgaard and William Sinton have read all or parts of the manuscript and have made valuable suggestions that are part of this book.

Artist Ray Lanterman took the story I wrote and represented it visually, while Lee Kyselka furnished an objectivity and wisdom that made the creative effort a published reality. Editor Elizabeth Bushnell sharpened the text, and Douglas B. T. Chun assisted with the visual presentation.

A word about the units of measurement. Distances are expressed freely in either the British or metric systems. A casual approach, using either or both, seems appropriate now that the United States is in a state of transition from the British system to the metric. A useful approximation of distance is this: 1 km = 0.6 mile; and, conversely, 1 mi = 1.6 km.

W. K.
Honolulu, Hawaii

# ONE-FIFTH OF A TURN

Astronomy came out of the past. Pioneer 10 is sailing into the future—the first artifact ever to leave the gravitational influence of the Sun.

Two million years from now it will be near Aldebaran. Two million years is time enough for light coming from the Andromeda Galaxy to reach Earth, time enough to carve a Grand Canyon, and time enough to add another volcanic island to the Hawaiian Chain.

Early man may have been living on Earth two million years ago. Since that time, the Earth's polar axis has traced some 80 precessional circles on the celestial sphere. But during the last fifth of a cycle, man's way of living changed dramatically as he moved out of caves and began farming, building cities, and writing on clay tablets.

It is only a fifth of a precessional turn from Pyramids to Pioneer 10, from megaliths to miniaturization, from Stonehenge to silicon wafers. Only a fifth of a turn from the Tower of Babel to the gantry tower, from the invention of the wheel to the abandoning of automobiles on the Moon. Only a fifth of a turn from the tilling of the soil to the sampling of the surface of Mars in search of life elsewhere in the universe.

The Sun has made 20 trips around the center of the Galaxy. Now in middle age, it has yet another 20 trips to complete before swelling into a red giant that will engulf the Earth. Spacecraft now leaving the Sun's gravitational influence will escape that ultimate catastrophe. Other craft will join them. And our knowledge of the universe will increase as they move outward from the Sun, carrying scientific instruments silently toward stars as yet unborn.

# FROM OUT OF THE PAST

*The soft, mystical brilliance of the*
*Eastern night.*
*The impelling beauty of the stars,*
*demanding veneration.*
*And Seven Spheres*
*moving through the heavens*
*with god-like serenity:*
*Surely,*
*in such celestial majesty*
*and harmony*
*there must be a Guiding Providence!*

The Babylonians were moved to deep religious feeling for the heavens. Lion, Bull, and Fishes; Twins, Ram, and Scorpion—each constellation had its myth, each was worshipped for its powers, and each was a fixed and eternal part of the sky.

But even more important than the stars that formed the constellations were the Seven Wanderers—Saturn, Jupiter, Mars, the Sun, Venus, Mercury, and the Moon—the very gods themselves manifest in the heavens, now gathering in one constella-

tion, then dispersing to reassemble in another. So powerful were these celestial wanderers that they influenced all the events on earth. Surely, man's destiny *must* be related to the stars!

If man could only know the meaning in the movement of the spheres, he would then be able to predict all the events on earth.

The Babylonians were careful observers. Studying the stars was to them an act of divine worship, a searching-out of the will of the gods as expressed in the heavens. Chief among the searchers was the astronomer-priest who knew the way of the gods—a powerful person to whom the king submitted humbly for guidance in directing the affairs of state.

Both astrology and astronomy began as observational sciences. Later they parted company, astrology moving toward mystical speculation. Basic to ancient astrology was the idea that planets, being of divine nature, influence human destiny. So it became a search for the correlation between the happenings in the heavens and events on earth. Astronomy deals with the measurement of sizes, distances, chemical compositions, and motions of stars and galaxies, and with the determination of the origins of the universe.

The Greeks borrowed from Babylonian astrology, shaping it to their own culture. No longer was the horoscope reserved for royalty alone; it was expanded to include everyone. The emphasis shifted from predicting collective destiny (the king as leader of the state) to individual destinies. Each person had a horoscope. His fate, readable in the stars, was fixed by the unique celestial configurations at the moment of his birth (although some argued that it should be at the moment of conception).

Divination became the dominant mode of astrology—the zodiacal circle becoming the center of a geometry of fate so complicated, so intricately interwoven with balancing, negating, and criss-crossing influences, as to become virtually a science of its own.

The Greeks secularized the Babylonian heavens, creating myths to go with star groups yet unnamed and filling the firmament with gods and heroes, now known by their Latin names—Perseus, Andromeda, Pegasus, Orion, Hercules. No longer worshipped as a whole, the heavens became a patchwork array of sacred objects, sidereal monsters, animals, and heroes. The Greeks, then, added human qualities to the heavens in constellation figures embodying arrogance, love, fear, strength, compassion, vanity, and tenderness.

## MYTH AND MATH

Both astronomy and astrology flourished in the intellectual atmosphere of Greece. Developing alongside a mythological worldview was one based on geometrical form and mathematical relationship. The lure of celestial perfection as expressed in number and form was so strong, though, that simple observation was slighted.

Anaximander, about 600 B.C., put forth the idea that the sun is a sphere 28 times the size of the earth, hollow and full of fire, and that stars are compressed air emitting flames from nozzles. Based on familiar observable mechanisms, his approach was a great step forward toward a rational view of the universe.

Eudoxus imagined a universe of spherical shells nesting one within another, each containing a planet. Pythagoras, a mathematician with a flair for the mystical, had everything in the universe moving in concert and harmony. Nothing moved randomly. Mathematics described that "harmony of the spheres," just as it defined in music the octave, fifth, and third as number relationships on a vibrating string.

Science began with the Greeks applying mathematics to the study of the heavens. By the middle of the third century B.C., Aristarchus had calculated the distances to sun and moon. His lunar distance was good, but his solar distance was off by a factor of 20, simply because precise measuring instruments were not available at that time. A generation later, Eratosthenes determined the earth's circumference. His figure was correct to within 250 miles of the modern value, depending upon the value assigned to his unit of measure, the *stadia.*

The search for meaning in the movement of the visible gods in the heavens led man to an understanding of the motions of the planets. Gradually it appeared that regularity, not divine whim, governed their movements. Most often, the Bright Wanderers traveled slowly eastward against the background of zodiacal stars. Now and then, however, they would reverse (except the sun and moon), moving toward the west for a while in *retrograde* motion, only to reverse again and to start traveling eastward in *normal* motion. Where there's regularity, there's predictability.

Predictability led to further mathematical models of the heavens. Spheres-within-spheres, rotating one within another, were invented to explain the vexing problem of variations in planetary speeds. Aristotle increased the spheres to half a hundred; still he

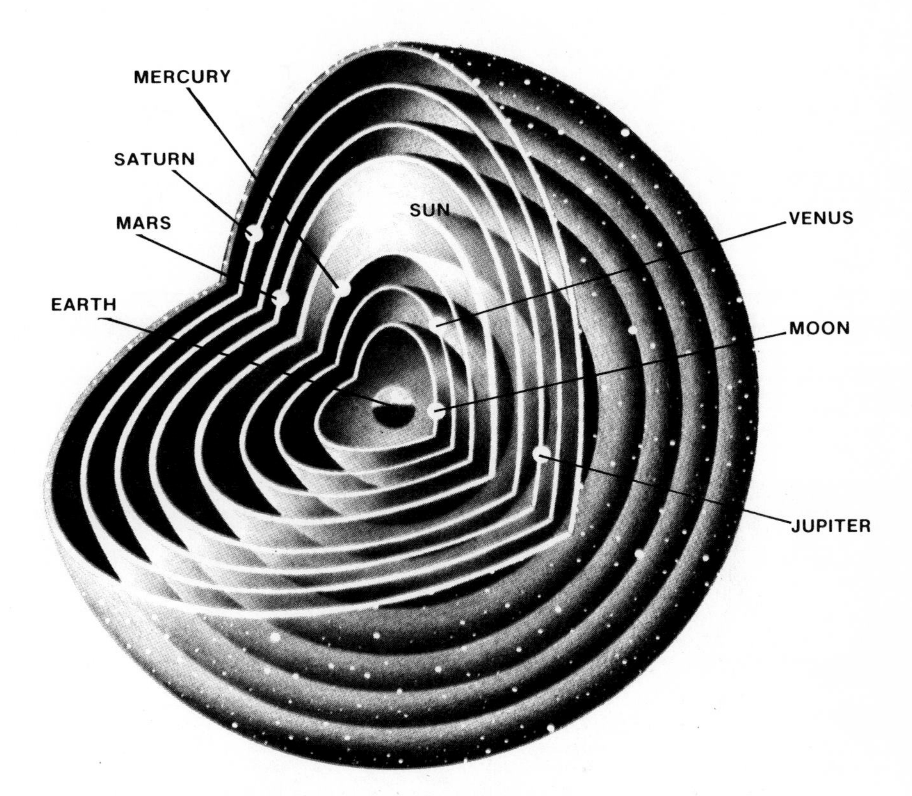


**The universe of Pythagoras. Sun, moon, and planets each in its own celestial sphere, moved around the earth. Surrounding all was the sphere of fixed stars.**

could not establish their harmony. Something was always off by just a little—and that led to models of still greater complexity.

The understanding of the motions of the planets eluded reason alone—an impasse that only the accumulation of better data could hope to break.

Hipparchus was the man to do it. A data-gatherer more than a philosopher, he invented instruments and charted the positions of a thousand stars. His instruments were so accurate and his technique so refined that by 150 B.C. he had determined the length of the year accurately to within 6 minutes of its true value. Three centuries later, Ptolemy relied almost entirely on Hipparchus' data as the basis of his world system.

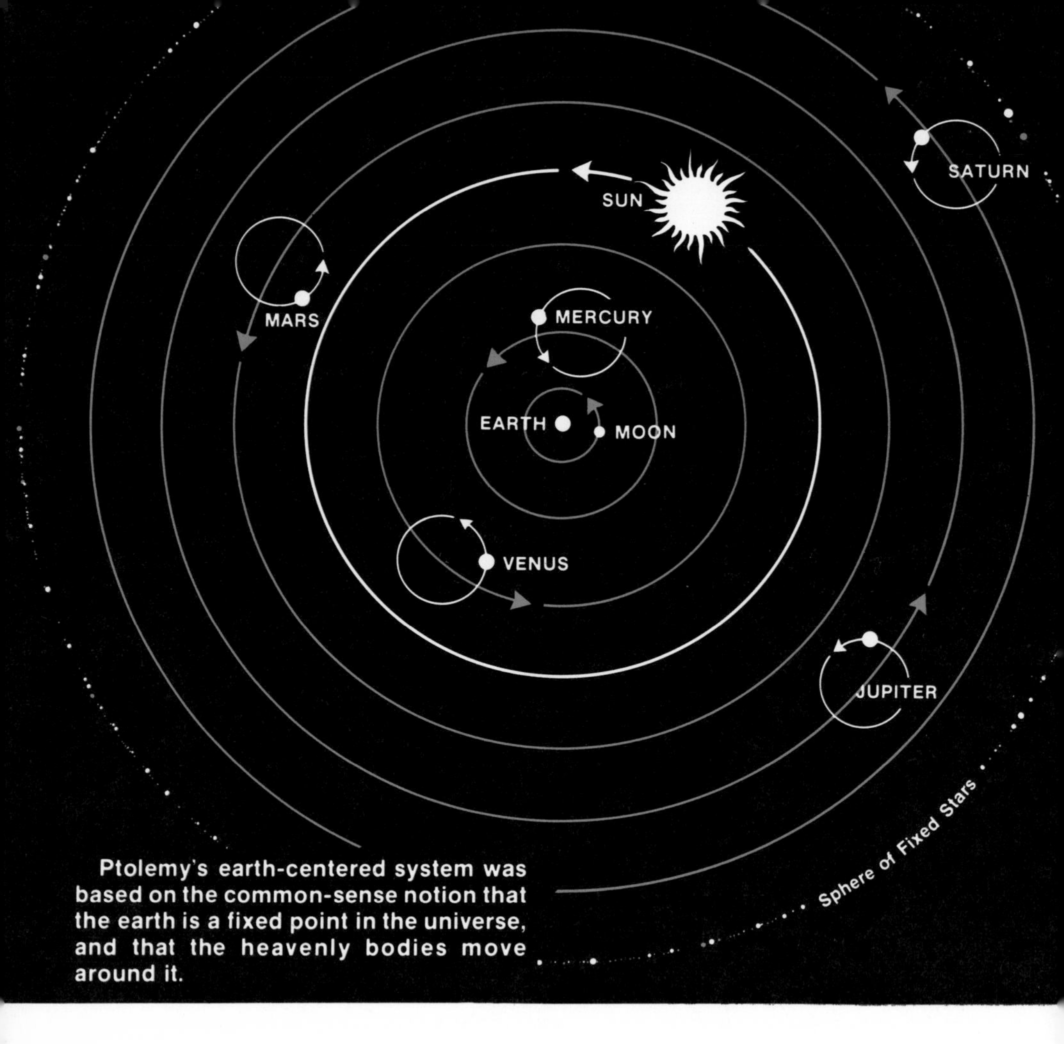

Ptolemy's earth-centered system was based on the common-sense notion that the earth is a fixed point in the universe, and that the heavenly bodies move around it.

Ptolemy constructed a model of the universe that could explain the retrograde movements of the planets as well as the variations in their speeds and brightnesses. Planets move in little circles, *epicycles,* as they travel in larger circles, *deferents,* around the earth. Ptolemy did not invent the epicycle-deferent model, but he did bring it to its elaborate and accurate state.

Circles upon circles—*epicycles upon deferents*—how complicated the celestial clockwork! But it worked—almost perfectly. More difficult to accept than its ten-thousand motions, though, was the fact that the earth was not exactly at the center of it all, but just slightly to one side. How vexing the eccentric earth when one is looking for the perfection of the centered circle!

Cumbersome as it was, the Ptolemaic model of the universe dominated Western thought for over 1,400 years.

## OTHER PLACES, OTHER PEOPLE

While astrology and astronomy were developing in Chaldea and Greece, other cultures had their own star-gazers who made discoveries and developed unique systems for understanding the universe.

### Egypt

The Egyptians, driven by the necessity of re-establishing boundaries after the annual flooding of the Nile, developed the art of surveying to a high degree. So accurately did they set the Pyramids to the cardinal points of direction that a recently discovered deviation of only 1/12 degree in the alignment of one has caused some scientists to wonder if it isn't more easily explained as Africa's participation in "continental drift" than as an error in Egyptian engineering. An inclined shaft in the Great Pyramid opened to the star Thuban in Draco the Dragon—the north star at that time. Their calendar was based on the emergence of Sirius from the sun's glow.

### Megaliths in England

While the Egyptians were building their Pyramids along the Nile, the Wessex people in southern England were hauling the biggest stones that ancient man has ever moved to build their observatories. The most famous of these megaliths, Stonehenge, was started about 2500 B.C., and it took centuries to complete. Hauling 30-ton blocks of stone from 20 miles to the north, they built a circle of uprights, capping them with lintel stones, which they bashed and shaped to fit the curvature. Within this *sarsen circle* are five narrow archways of 50-ton blocks that may have served as sighting portals to astronomical events on the horizon.

For more than 4,000 years, Stonehenge stood lonely and mysterious on the Salisbury Plain, not seeming to connect with anything in history. Caesar and his legions marveled at the structure 20 centuries after it was completed. No one then knew its origin.

Some speculated that the Druids built it for their rites and rituals, and others supposed that Merlin the Magician set it there. But modern archaeology shows that Stonehenge existed many centuries before the time of either Merlin or the Druids.

One fact was well known: an observer standing at the center of Stonehenge on the first day of summer could look down the main corridor and over a distant marker stone to the rising place of the sun. Other than that, no one knew its capabilities.

Not until a modern computer was used to answer the question, "Are there any other significant alignments?" was it known that the monument itself was a Stone Age computer. Its avenues and sighting portals are all aligned to seasonal rising and setting places of the sun and moon only, and are not significant with regard to the planets and stars. Built into the structure is a large circular lunar calendar. But most unusual at Stonehenge is a system for keeping track of the 18.6-year *nodal regression* of the moon, giving it the capability of predicting eclipses.

No written messages are inscribed at Stonehenge as there are in the Pyramids. But in the silence of the huge stone structure on the Salisbury Plain, the mind of archaic man speaks clearly, showing that he was capable of highly abstract thought long before he invented the art of writing.

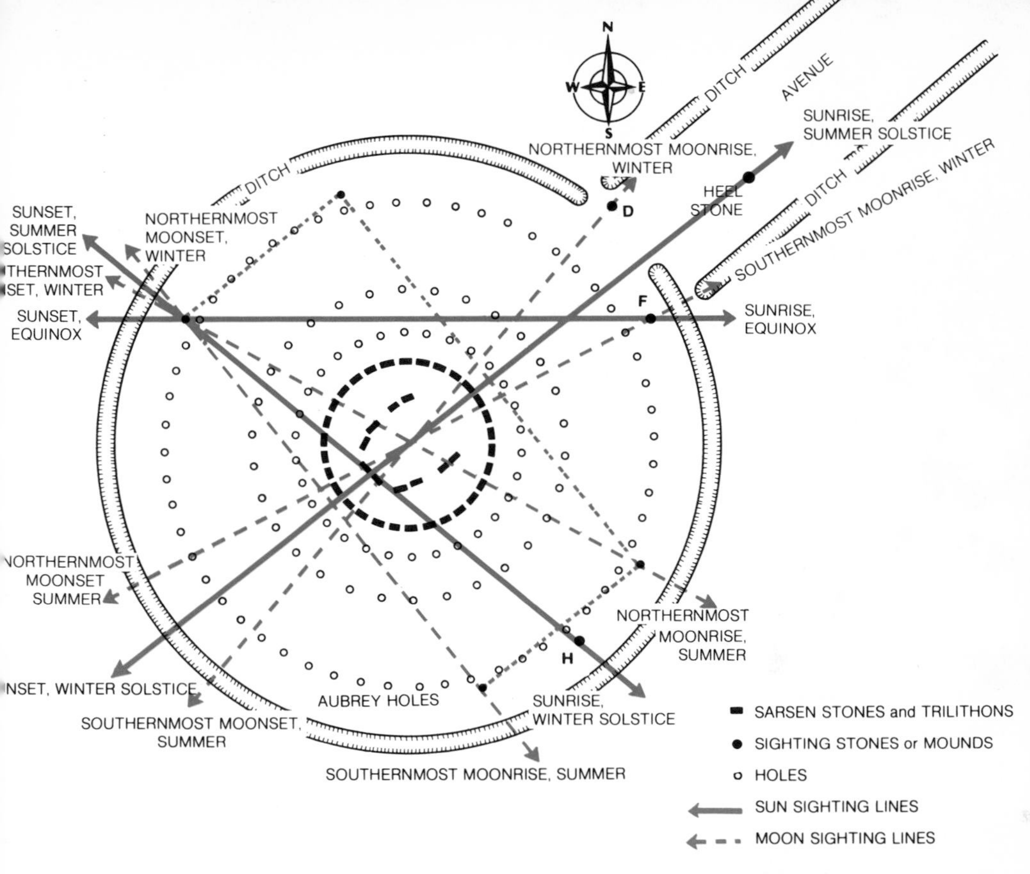


**Opposite page: As the sun rises at summer solstice, the Heel Stone casts a long shadow down the main avenue of Stonehenge.**

**Above: The 400-foot circular ditch was dug at Stonehenge about 45 centuries ago. Two or three centuries later, huge stones were dragged great distances, then raised and capped with lintel stones. The U-shaped arrangement of 50-ton blocks opens toward the mid-summer rising position of the sun.**

**Some of the significant astronomical alignments are shown in blue.**

## China

On the opposite side of the earth, the Chinese were keeping excellent astronomical records, and by 2000 B.C. had determined the length of the year to be 365 days. Determining the cardinal points of direction to a high degree of precision, the Chinese oriented their buildings to those points and annually reaffirmed that knowledge in elaborate ritual.

**Chinese scholar at work before a 5th century circumpolar sky map showing the Big Dipper.**

## Central and South America

The Mayan people of Central America, too, determined the length of the year, developing a calendar that enabled them to understand better the productive cycle. And they left temples oriented to the stars.

Markings on the tops of the arid plateaus of Nazca, south of Lima, Peru, have long been a mystery of ancient America. Lines, from a few meters in length up to several miles, run in all directions, then end abruptly. Overlapping rectangles, trapezoids, and triangles are present, mixed with gigantic 400-foot figures of birds, fishes, monkeys, and spiders. Recent archaeological work has shown that some of the lines run toward the position of the rising sun at the times of the solstices, suggesting that this "astronomy book" may have been used for predicting the changes in seasons and so indicating when water might be available in this desert region.

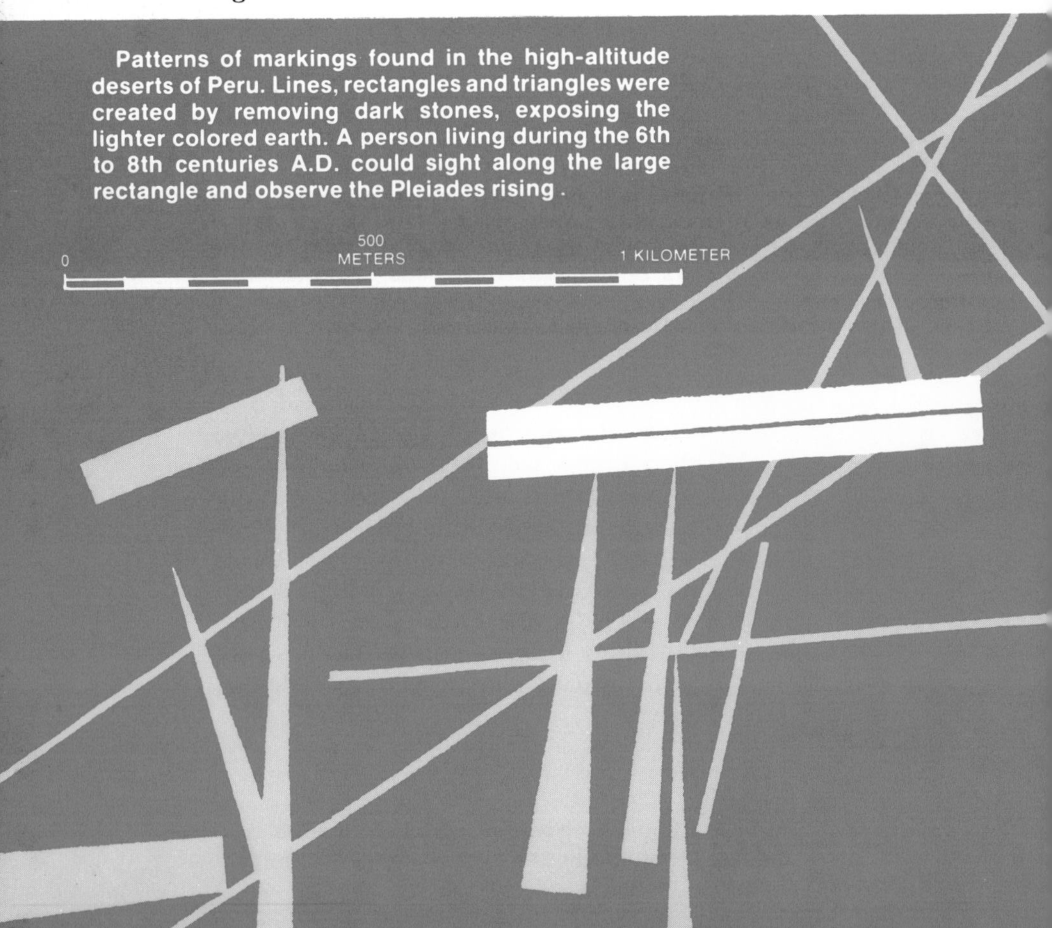


Patterns of markings found in the high-altitude deserts of Peru. Lines, rectangles and triangles were created by removing dark stones, exposing the lighter colored earth. A person living during the 6th to 8th centuries A.D. could sight along the large rectangle and observe the Pleiades rising.

## The Pacific

The Polynesians made practical use of the stars in voyaging all over the Pacific Ocean. Sailing from beneath the brightest star, *'A'a,* glowing (now Sirius), they reached tiny islands over thousands of miles of open sea and returned, all without the use of navigational instruments. Chants, learned letter-perfect and transmitted from one generation to another, contained the sailing instructions. Sustained by a worldview that was not hostile toward man, and with a knowledge of the ways of the sea inherited from generations of sea-faring people, they sailed forth on one of man's greatest ventures with only the light of stars thousands of billions of miles distant to guide them over this watery planet.

## SWITCH IN MODELS—GEOCENTRIC TO HELIOCENTRIC

The Ptolemaic system stood for 1,400 years. It represented straight thinking—the concept of the earth as the center of the universe. It worked well for a while. But small deviations shifted to great extremes over the centuries. Alfonso expressed his dissatisfaction with the system a thousand years after Ptolemy with a great irreverence: "Had I been present at the Creation, I would have given some useful hints for the better ordering of the universe."

Copernicus, in 1543, put the sun at the center of his world model. The idea had appeal, for it neatly explained retrograde motion as an apparent one—a planet appearing to move backward against the background of stars as the earth passes it. Aristarchus had originated the sun-centered model long before Copernicus, but the idea was dropped when it was found that its uniform circular orbits did not represent the observations well.

The Copernican system was hard for many people to accept, for it violated common sense. Earth movement around the sun must be so fast that a whole city would fly by in a second. How could we possibly partake of such motion and not feel it? A complicated system of epicycles and deferents was still required to make the Copernican system work; so it was not any simpler, nor was it any more accurate than the Ptolemaic system.

A century passed before the heliocentric universe was generally accepted. It takes 50 to 100 years for such a revolution in thought to take place, just as in the 20th century it has taken 50 years for "continental drift" to be accepted. In both revolutions, it was the accumulation of data plus a mathematical framework that made revolutionary ideas plausible. The mathematical work of both Kepler and Newton clinched the debate on the sun-centered over the earth-centered model of the universe. Kepler found that the planets actually move in ellipses, not circles, and Newton explained why.

Scientific instruments invented in the 16th century increased man's ability to gather data. Some of the instruments became immortalized in the sky as new knowledge led to new cosmologies. Constellations named most recently, about 200 years ago, include

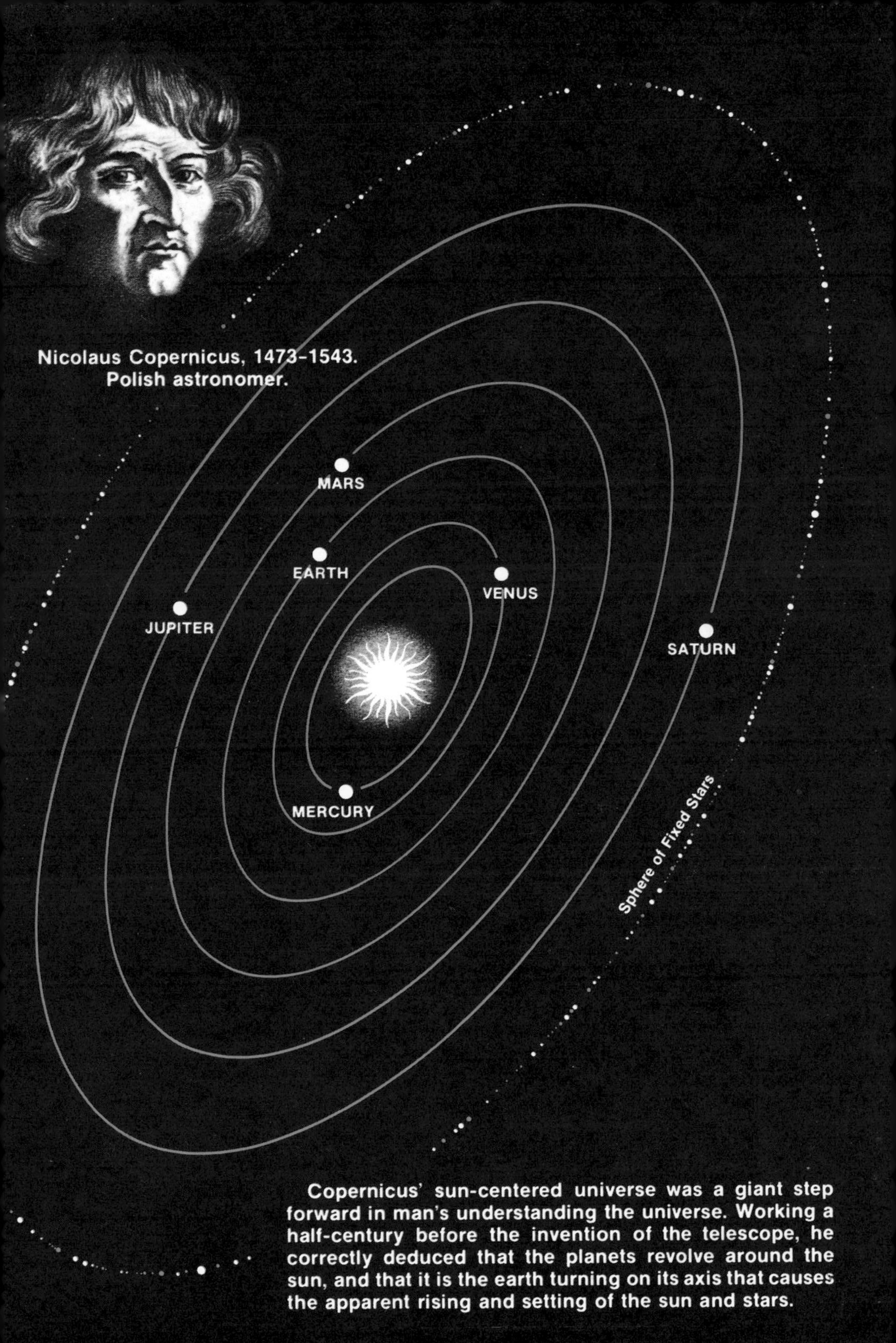


Copernicus' sun-centered universe was a giant step forward in man's understanding the universe. Working a half-century before the invention of the telescope, he correctly deduced that the planets revolve around the sun, and that it is the earth turning on its axis that causes the apparent rising and setting of the sun and stars.

an air pump, a chemist's furnace, a sextant, a pair of compasses, a microscope, and a telescope.

Thousands of years ago the heavens made sense to the Babylonians as the dwelling places of the gods. But as undesignated areas of faint stars were imagined as monsters, heroes, and scientific instruments, the sky became confusion confounded. No longer does it make sense as a whole, even though it contains a colorful array of the thoughts of mankind. It makes no sense that a chemist's furnace and a clock are on opposite banks of the celestial river Eridanus into which Phaëton fell. Fanciful though they may be, constellations are useful in establishing place names in the sky for ease in locating celestial objects. Now there are 88 recognized consellations, each situated within an area established by international agreement.

Star patterns that the ancients saw are the same ones we see today, little changed over the millennia. Stars *do* move, though, some at speeds of hundreds of kilometers per second, so that constellations are changing slowly. But stars are so tremendously far away that the 50 centuries during which man has been observing the sky is too short a time for him to see a change.

Star patterns have not changed, but man's relationship to them has. The impelling beauty and majesty of the cosmos that led ancient man to worship the heavens moves us, too. We've lost the feeling for myth, but our knowledge takes us far beyond the veneration of shining orbs, and we've photographed distant galaxies whose light has taken billions of years to reach us.

So we, too, stand at the edge of the abyss and, with a feeling for the immensity and eternity of the universe, share Pascal's feeling, "The eternal silence of these boundless spaces frightens me."

Sunrise over the Heel Stone at Stonehenge as it may have appeared in ancient times when lintel stones capped the uprights in the sarsen circle.

North American Indians arranged wooden posts—"Woodhenges"—oriented to the equinoctial and solstitial positions.

Below: The Mayans of Central America built temples oriented in a similar manner.

# THE ZODIAC

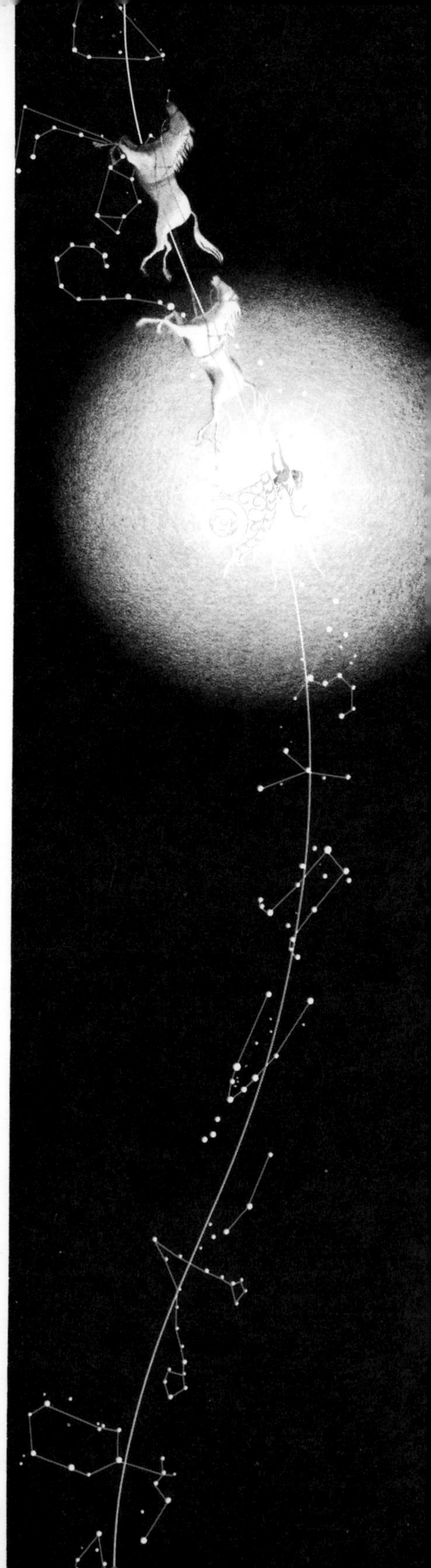

*Man, in his attempt*
*to understand the mystery in the*
*movements*
*of the sun, moon, and planets,*
*endowed*
*twelve star groups along the*
*path of the sun*
*with power.*

*Born in mysticism,*
*the age-old zodiac is*
*a useful frame of reference for the*
*astronomer*
*as he explores the depths of space*
*far beyond the stars*
*that ancient man*
*could see.*

During the year, the sun moves slowly eastward through twelve famous star groups—the "Circle of Animals" or *zodiac.*

Study of the sun's relationship to the stars along its path gradually led man to an understanding of the motions of the earth—the spinning that brings us night and day, the journey around the sun that gives us the seasons, and the wobbling that accounts for a succession of pole stars.

The zodiac is a belt 16 degrees wide running across the sky, 8 degrees on either side of the path of the sun. Moon and planets, varying only slightly from the sun's path, also move through the zodiac; and the band is wide enough to accommodate the greatest deviations of these Wanderers.

Constellations in the zodiacal band vary in form and clarity. Taurus, Gemini, Leo, Scorpius, and Sagittarius are bright and easy to find. Virgo, Libra, and Aries have a single bright star or two but are not clear in outline. Cancer, Capricornus, Aquarius, and Pisces are faint and hard to see even under the best of conditions. Regardless of form or brilliance, each is important—to the astronomer as points of reference for locating celestial objects; to the astrologer for the mystical character with which each has been endowed.

Zodiacal figures embody ancient ideas. Ram, Bull, and Lion represent power. The fish-tailed Sea-Goat and the man-horse Archer are fanciful figures combining the best of two worlds. Crab and Scorpion are creatures who subdued the giants Hercules and Orion. The Fishes symbolize the rainy season. Virgin, Water Carrier, and Twins are the human forms in the Circle of Animals. The only non-zoological zodiacal group is Libra, the Scales.

## ALONG THE SUN'S PATH

The discovery of the sun's path through the heavens—the *ecliptic*—and the development of a zodiac may be among the first astronomical achievements of any culture. The first complete representation of the zodiac is the Atlante Farnese celestial globe, a Roman sculpture of 200 B.C. The Egyptians may have had a zodiac as early as 5000 B.C., for the zodiac on the dome of the temple at Dendra, although built in A.D. 100, begins with the Twins.

The Chinese established their zodiac far back in antiquity. They knew the paths of both sun and moon well enough to be able to predict eclipses. Each constellation in the *Kung* (zodiac) is an animal:

The Tiger (Sagittarius)
The Hare (Scorpius)
The Dragon (Libra)
The Serpent (Virgo)
The Horse (Leo)
The Ram (Cancer)
The Ape (Gemini)
The Cock (Taurus)
The Dog (Aries)
The Boar (Pisces)
The Rat (Aquarius)
The Ox (Capricornus)

Progressing in reverse order from our own, the Chinese zodiac has given rise to a cycle of years—the Year of the Tiger, the Year of the Hare, and so on around the circle.

New zodiacal figures have been invented, but ancient ones still prevail. The Venerable Bede turned them into Twelve Apostles in A.D. 700. A thousand years later, Sir William Drummond formed them into a dozen Biblical Patriarchs; and the classical scholar, the Reverend G. Townsend, changed them into Twelve Caesars.

Twelve became a magical number and a powerful symbol—twelve jurors, twelve segments in rose windows of cathedrals, twelve figures on walls, in paintings, and on pavements in both public and private buildings. Twelve seemed to be in tune with the universe—the moon sweeping twelve times around the earth with twelve days left over as the sun traveled through the twelve zodiacal constellations.

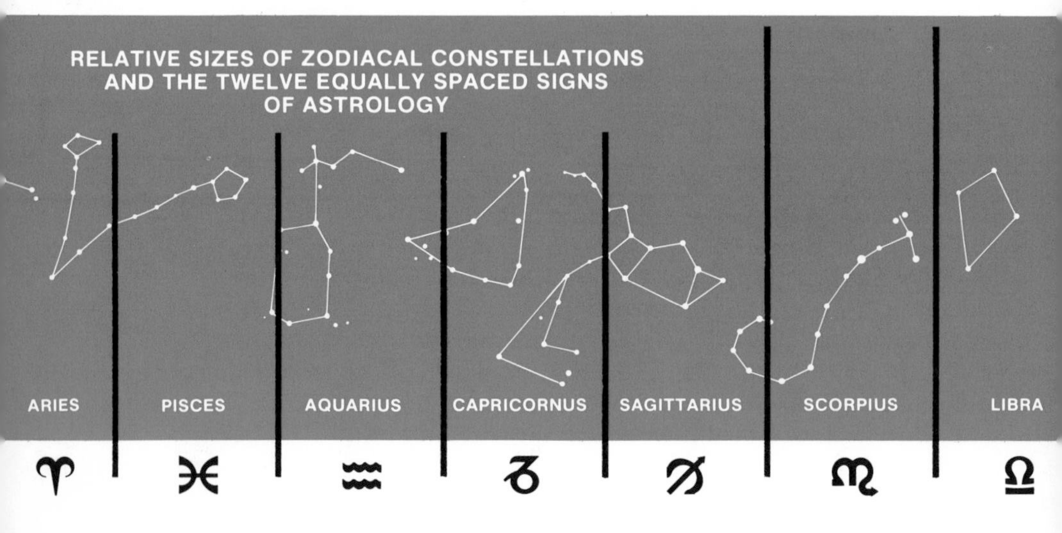

Zodiacal constellations vary greatly in size—a problem for ancient astrologers. Virgo is three times as big as Aries; Pisces, twice the size of Cancer. Just where does one group end and the next one begin? When is Jupiter no longer in Capricornus but in Aquarius? And at what point does the sun leave Taurus' influence for that of the Twins? The astrologer solved this problem of indefinite and irregular boundaries simply, mathematically, by dividing the zodiac into twelve equal parts, or *signs*. Each segment measured 30 degrees. And the twelve signs governed as follows:

♈ **Aries,** the Ram: March 21–April 19
♉ **Taurus,** the Bull: April 20–May 20
♊ **Gemini,** the Twins: May 21–June 21
♋ **Cancer,** the Crab: June 22–July 22
♌ **Leo,** the Lion: July 23–August 22
♍ **Virgo,** the Virgin: August 23–September 22
♎ **Libra,** the Scales: September 23–October 23
♏ **Scorpius,** the Scorpion: October 24–November 22
♐ **Sagittarius,** the Archer: November 23–December 21
♑ **Capricornus,** the Goat: December 22–January 19
♒ **Aquarius,** the Water Carrier: January 20–February 18
♓ **Pisces,** the Fishes: February 19–March 20

The astronomer still uses the ancient constellations, not the signs, as a convenient frame of reference in his observations.

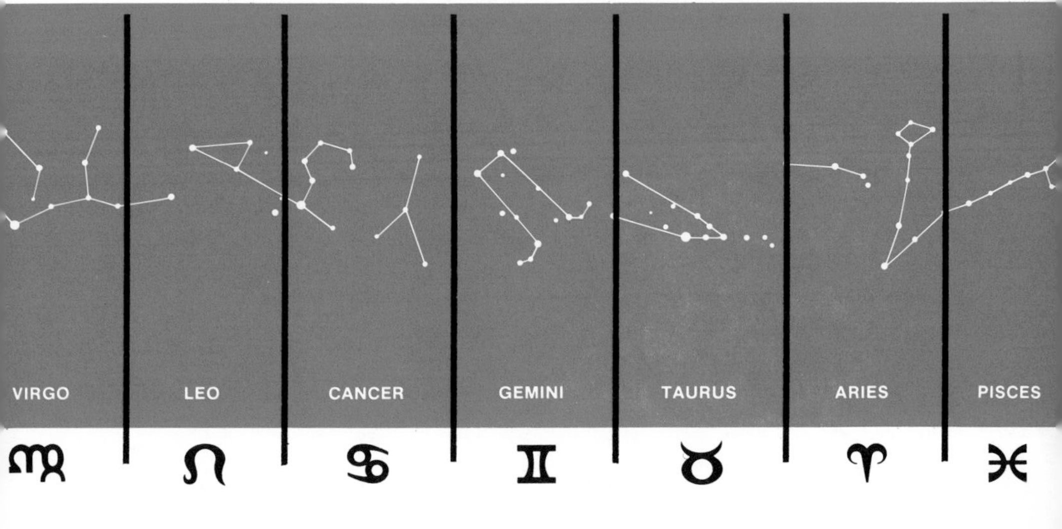

The astrologer went on to divide the sky into twelve "houses" —six below the horizon, six above. Sun, moon and planets were living beings, and these houses their dwelling places. Purely geometric, the idea of houses appeared at about the time astrology went to Greece. It was an earth-bound framework, a stationary system projected onto the celestial sphere. Constellations drifted westward through the framework during the night, visiting one house and then the next, spending two hours in each.

**Right: The twelve zodiacal houses. The first house is just below the eastern horizon.**

**Below: The horoscope of a person born at 6:30 A.M. on June 29, 1916, at 108° west longitude.**

**The sun ☉ in Cancer ♋ determines the person's "sign," and Leo ♌ is about to rise. Neptune ♆ is in the first house. Mercury ☿, in the eleventh house, is very close to Gemini ♊. Gathered together in the twelfth house are Pluto ♇, Venus ♀, Saturn ♄, moon ☾, and sun.**

**Various symbols in other parts of the diagram are interpreted as present-day influences.**

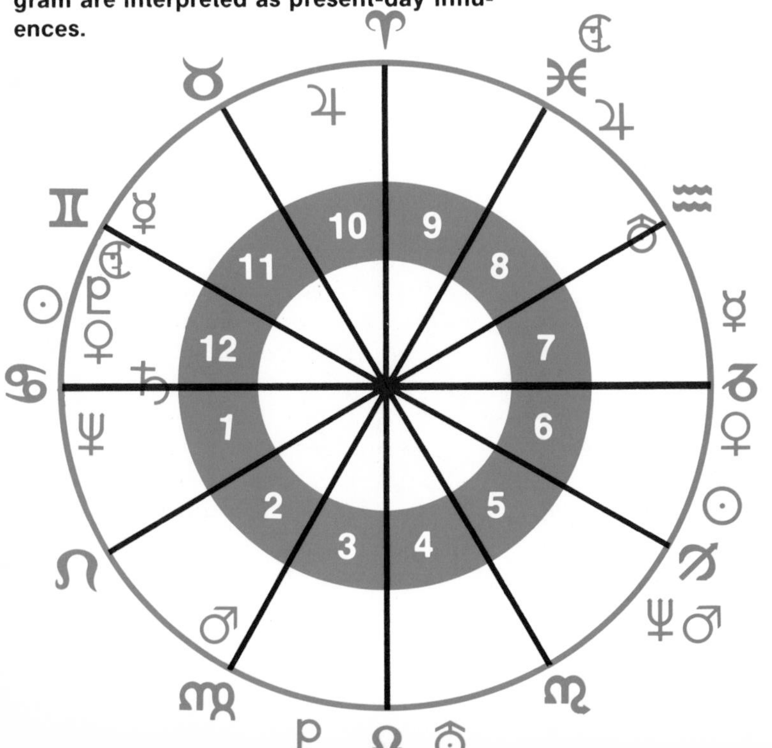

Twelve houses and twelve signs, with seven Wanderers moving independently through them, arranging and rearranging themselves in sextile, quartile, trine, and opposition positions—all these together becoming a blend of countless variables, related by geometry and number and requiring study and interpretation by skilled astrologers. So it was in Greece that mathematics coupled scientific achievement to speculative thought.

## CELESTIAL CIRCLES

Careful observation leads to the inference of two important circles in the sky—the ecliptic and the celestial equator.

Each day the sun rises in the east, travels in an arc across the sky, then sets in the west. *Daily* motion, due to the spinning earth, brings a regular variation in light intensity on our planet—a rhythm to which life is attuned. But each day the sun moves in a slightly different path, and so each day is of different length than the one before.

The sun has an *annual* motion, too, that carries it through the zodiac in a year. Each day it is a little farther to the east against the background of stars (about one degree) than it was the day before. The sky is much too bright for us to see those zodiacal constellations while the sun is there (except during total eclipses). But it is easy to tell where the sun has been and where it will be relative to the stars by watching at morning and evening twilight. Ahead of the sun in the morning and separating itself farther from it each day is the star group that the sun has just visited. Twinkling in the evening twilight above the setting sun are the stars of the zodiacal group into which the sun will soon move.

Each day the sun rises and sets in a different place along the horizon. The difference is slight—so slight that it takes several days of observing to notice much change. Late in December (in the northern hemisphere) the sun sets far to the southwest. All day it has been traveling low across the southern sky. Shadows at noon are long and point toward the north. And the sun seems to linger many winter days far in the south as if to gather forces for its six months' journey to the north. Late in June it makes its longest passage along the sky, setting in the northwest. The extreme positions of the sun are the *solstices (sol,* the sun + *sistere,* to stand still).

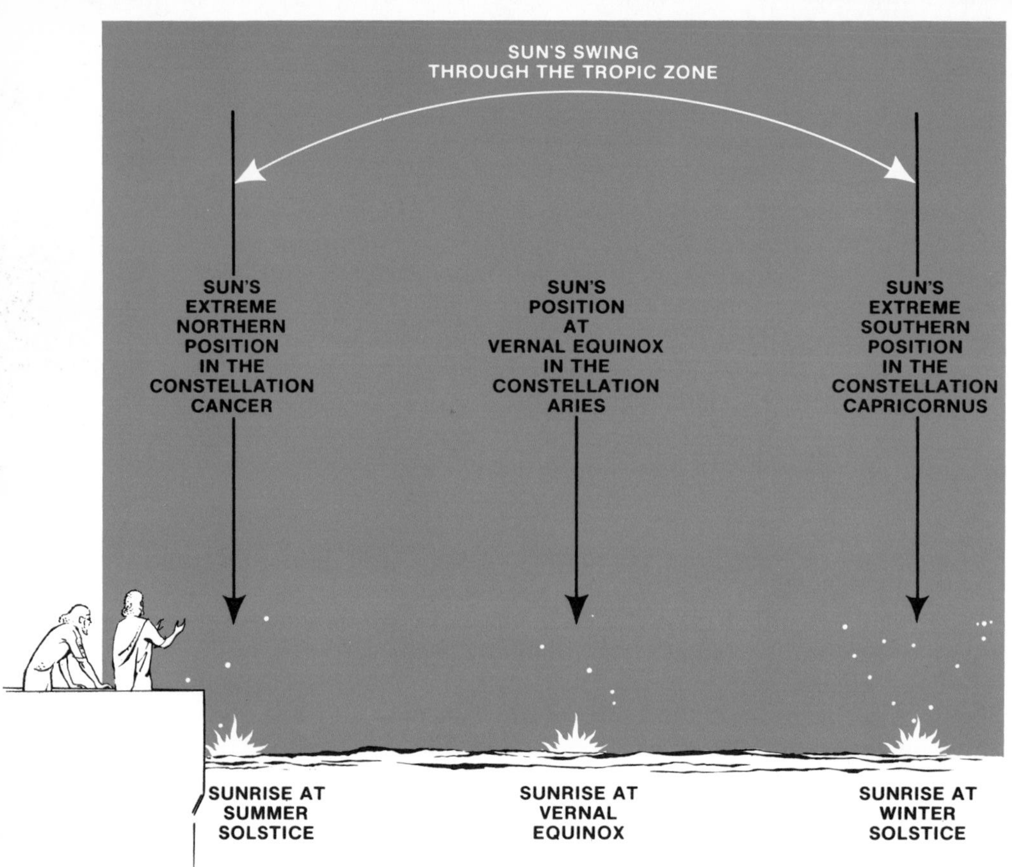


**The sun was in the direction of Aries at vernal equinox for the Babylonians. Three months later, when farthest north, it was in the direction of Cancer. Swinging southward, the sun crossed the equator in the direction of Libra, reaching the December solstice in Capricornus.**

Careful measurement shows that it takes a little more than 360 days for the sun to visit the two solstitial positions along the horizon and to return to its starting place. Here's a second way of determining the length of the year. The first was by measuring the time that it took the sun to move through the zodiacal constellations and get back to a given place among the stars; and now, the time that it takes the sun to move from its northern to its southern limit, and return.

The Babylonians knew that the stars of the Crab and Goat were beyond the sun at the solstices. (Gemini and Sagittarius are now there). In their times, the sun turned southward when in the direction of Cancer; northward, in Capricornus. The imaginary line on

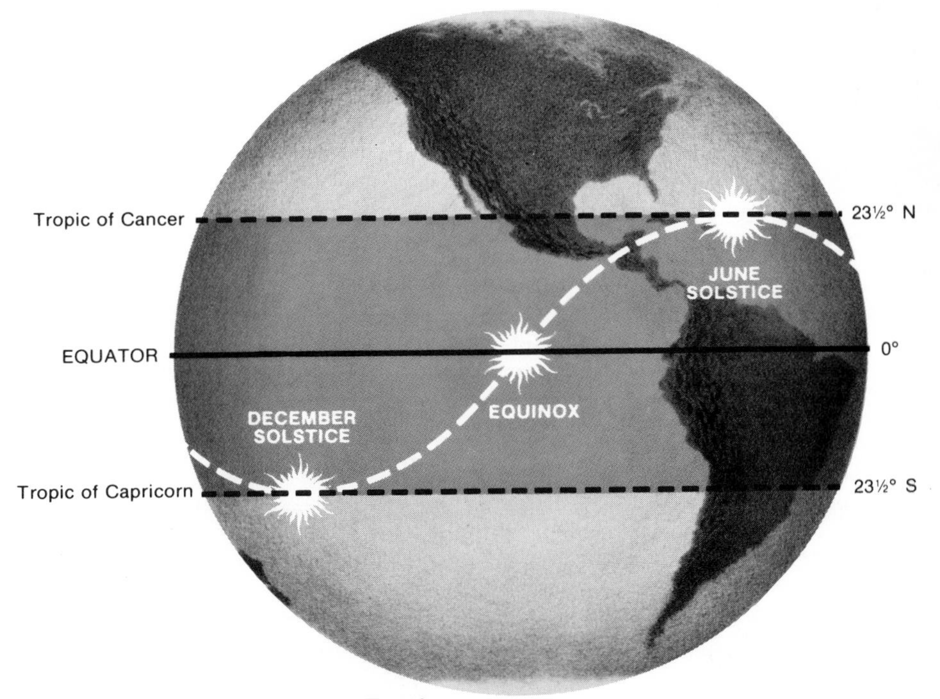


**The Sun travels from the Tropic of Cancer to the Tropic of Capricorn in six months. The region of Earth lying between the two is the "tropic zone"—23½ degrees on either side of the equator—shown here in blue.**

earth marking the northern limit is the *Tropic of Cancer,* "the turning place of the sun in the direction of Cancer." The southern limit is the *Tropic of Capricorn;* and the region between the turning-places of the sun, the *tropics.*

The sun crosses the equator March 21 on its swing into the northern hemisphere. This crossing, where ecliptic and equator meet, is the most important point along the ecliptic—the *vernal equinox.* All star positions are measured from it, and it is the base from which all astrology is reckoned. Changes are rapid. Winter gives way to spring with the consequent greening of the hemisphere. It's a season of renewal and a time for celebration—bacchanalian festivals, planting rituals, and the traditional rites of spring.

## SHIFTING POLES AND WOBBLING EARTH

Now the sun is in Pisces at vernal equinox. But it was not always so. The vernal equinox was in Aries when astrology began. Before that, in Taurus. And at the dawning of civilization it was in Gemini.

The westward drift of the vernal equinox through the zodiac is much too slow—only one degree every 72 years—to be detected by a single observer in a lifetime. So its discovery depended upon the accumulated observations of sky-watchers over a long period of time. Babylonians found the vernal equinox in the direction of Aries. Several centuries later, Hipparchus discovered that it was no longer at the same place in the Ram but that it had moved westward. He was at a loss to explain this *precession of the equinoxes*, and it was to be yet another 20 centuries before the reason could be known.

The reason is the wobbling earth. And with it comes a change in pole stars.

The earth, like a top, wobbles as it spins. Slowly, its polar axis traces a huge circle on the celestial sphere—a task requiring 26,000

years to complete. Polaris lies on that circle, and the earth's axis presently is pointing to within a degree of the "North Star." But the earth wobbles on, and in a few centuries there will be no north star. Eight thousand years from now the polar axis will be pointing close to Deneb in Cygnus; in 12,000 years, to a point near Vega. Each is brighter than Polaris (Vega is six times as bright),

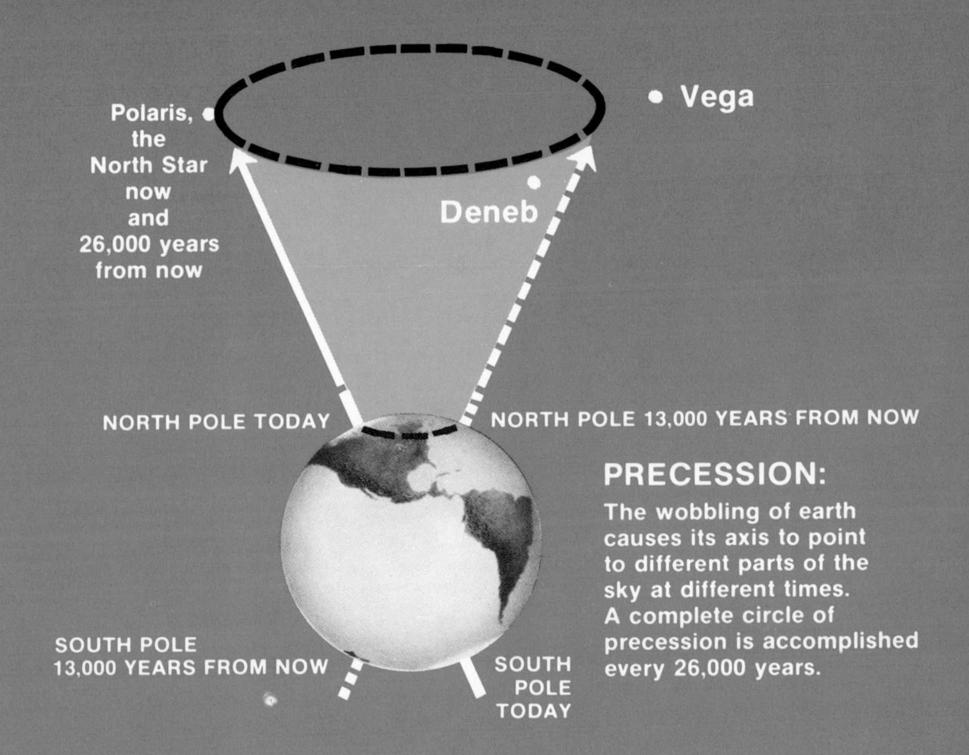


but each will be a less satisfactory north star, for each lies several degrees off the circle of axial precession.

Precession results in a gradual change in the stars that we associate with the seasons. Spring comes 20 minutes earlier each year relative to the stars as the vernal equinox moves along the ecliptic. Thirteen thousand years from now, Orion will be a summer constellation and Scorpius, a winter one. Precessional change is slow in human terms. Only one-fifth of a turn has been made since the building of the Pyramids. But 40 cycles are completed in a million years—in the time it takes to carve a Grand Canyon.

## PATH OF PRECESSION AMONG THE STARS ABOVE THE NORTH POLE

**The wobbling of earth causes an extension of its axis to trace a circle counter-clockwise among the stars over the northern polar regions. Tick marks on the circle indicate 1,000-year intervals.**

**A number of stars become "north stars" during the 26,000-year cycle of precession. Today we have a bright north star, for earth's north pole is pointing to within a degree of 2nd magnitude Polaris. Forty-five hundred years ago, 4th magnitude Thuban in Draco was the faint north star—the star toward which the Egyptians aligned central inclined shafts in their pyramids.**

**Two thousand years from now, the 3rd magnitude star at the head of Cepheus will be the North Star. Brilliant Deneb will be the pole star in A.D. 10,000, followed by blue Vega in A.D. 14,000. Though brighter than Polaris, Deneb and Vega will be much farther from the celestial pole than is our present-day Polaris.**

## PATH OF PRECESSION AMONG THE STARS ABOVE THE SOUTH POLE

**The earth's south pole traces a clockwise circle in the southern sky.**

**Today there is no "south star" as there is a north star. But between the years A.D. 7,000 and A.D. 14,000, many bright stars in Carina, Vela, and Puppis will qualify.**

**A line extending from Beta Centauri (Hadar) to Achernar now runs almost directly through the south celestial pole. Hadar is a tremendously brilliant star. Only slightly less brilliant to the eye than Alpha Centauri (Rigil Kentaurus), Hadar is a hundred times farther in space.**

**Both Alpha and Beta Centauri are 1st magnitude stars. The only other constellation having two 1st magnitude stars is Orion.**

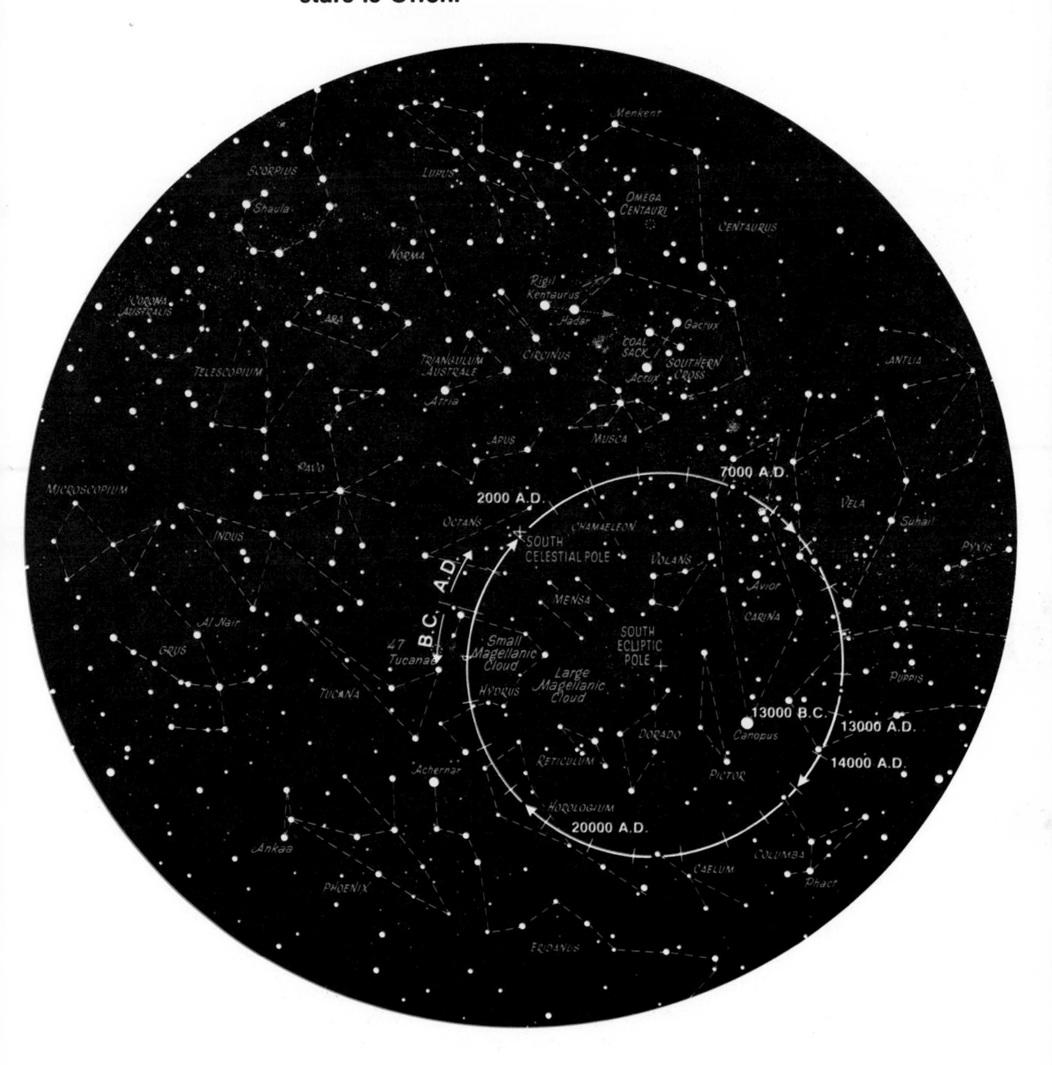

## ECLIPTIC AND EQUATOR

Ecliptic and equator are like two huge rings in the sky tilted 23½ degrees to each other. The rings intersect in two places, the *equinoctial points.* Each year the sun meets the equator at a slightly different point—a point that shifts slowly, completely through the zodiac in 26,000 years. If the equinox did not move, the year would be 20 minutes longer than it is.

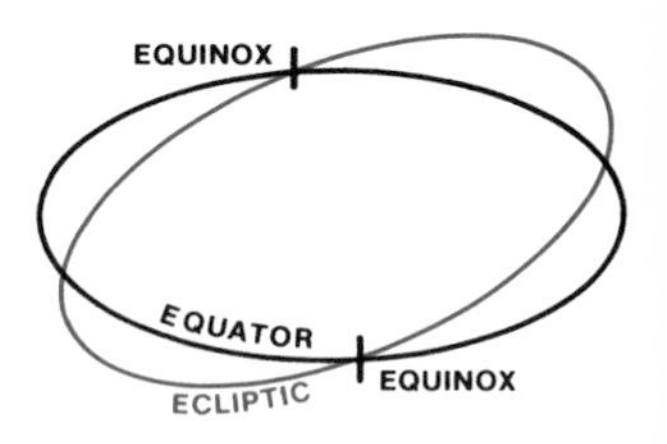


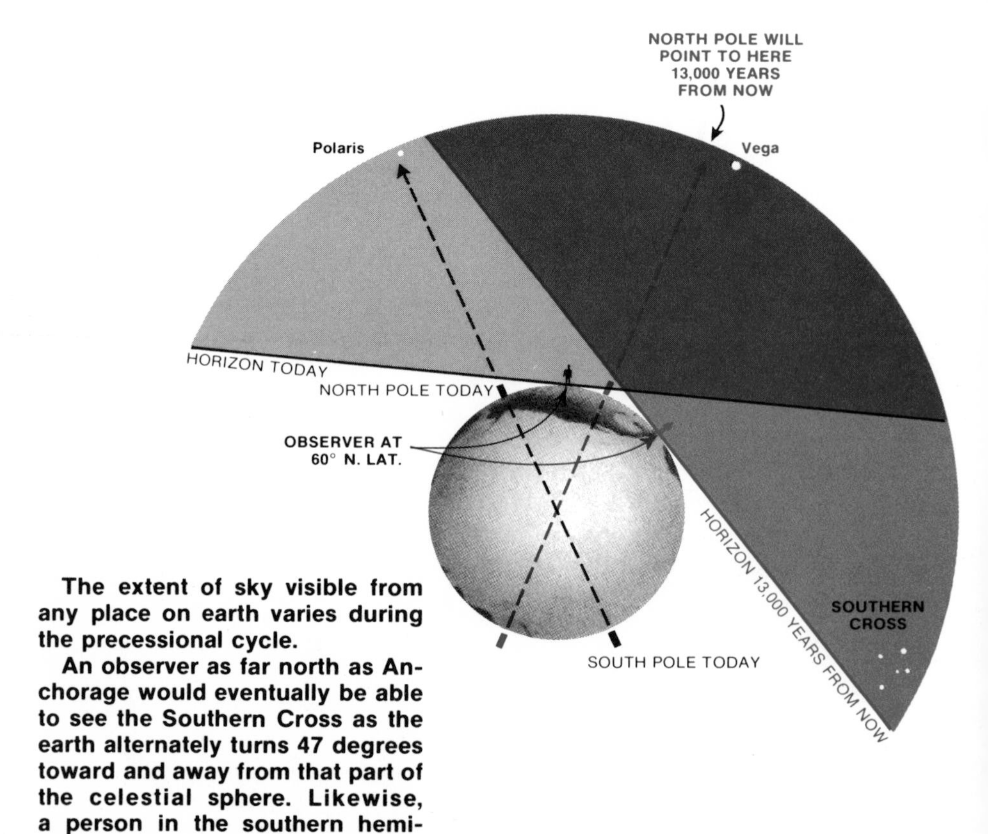


**The extent of sky visible from any place on earth varies during the precessional cycle.**

**An observer as far north as Anchorage would eventually be able to see the Southern Cross as the earth alternately turns 47 degrees toward and away from that part of the celestial sphere. Likewise, a person in the southern hemisphere today cannot see the north star, but Polaris would be visible to him at some future date.**

Precession results not only in changing pole stars and reversing seasons, but also in the appearance and disappearance of star groups from certain regions of the sky over long periods of time. The Southern Cross was visible from as far north as the Canadian border a few thousand years ago when Thuban in Draco was the north star. Now you must be as far south as the tropics in order to see it. More of the stars far to the south could be seen from North America in times past than can be seen now, and that continues to change over eons of time.

Precession results, too, in a confusion between celestial constellations and astrological signs. Once they were the same. No longer do they match. Vernal equinox, or "The First Point of Aries" occurred in the Ram 4,000 years ago at the beginning of the systematic observation of the heavens. Precession has since carried the vernal equinox westward into Pisces. Thus the "First Point of Aries" is now in the direction of Pisces. The sign of the Bull is now in the direction of the constellation Aries, and the sign of the Twins is in the direction of the constellation Taurus. No longer is the sun in Cancer on the first day of summer, but in the Twins; precession has made the Tropic of Cancer a misnomer in our day, for the sun is presently in Gemini when it is farthest north of the equator.

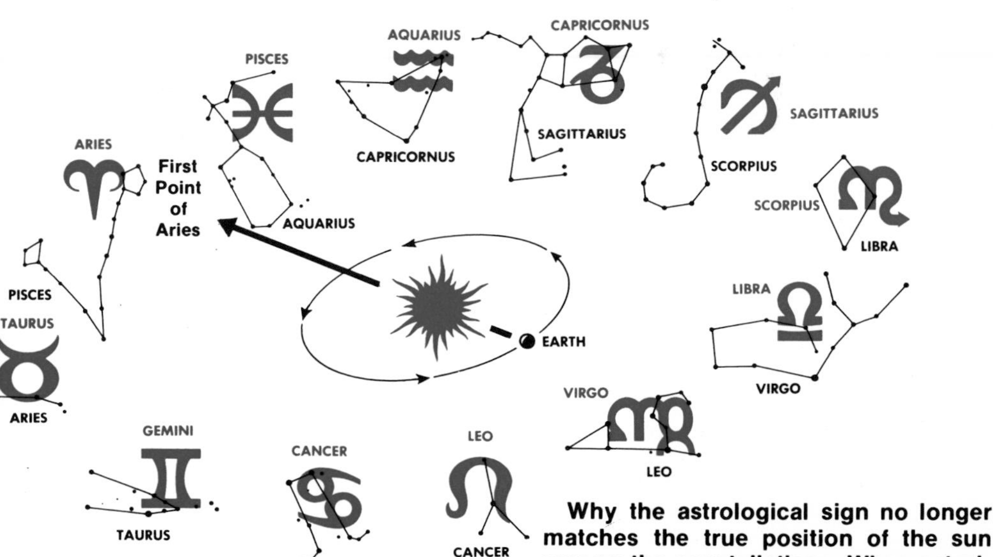


**Why the astrological sign no longer matches the true position of the sun among the constellations. When astrology began, the sun was in Aries at vernal equinox, but due to precession it has shifted westward. The First Point of Aries is now in the direction of Pisces, and will soon move into Aquarius.**

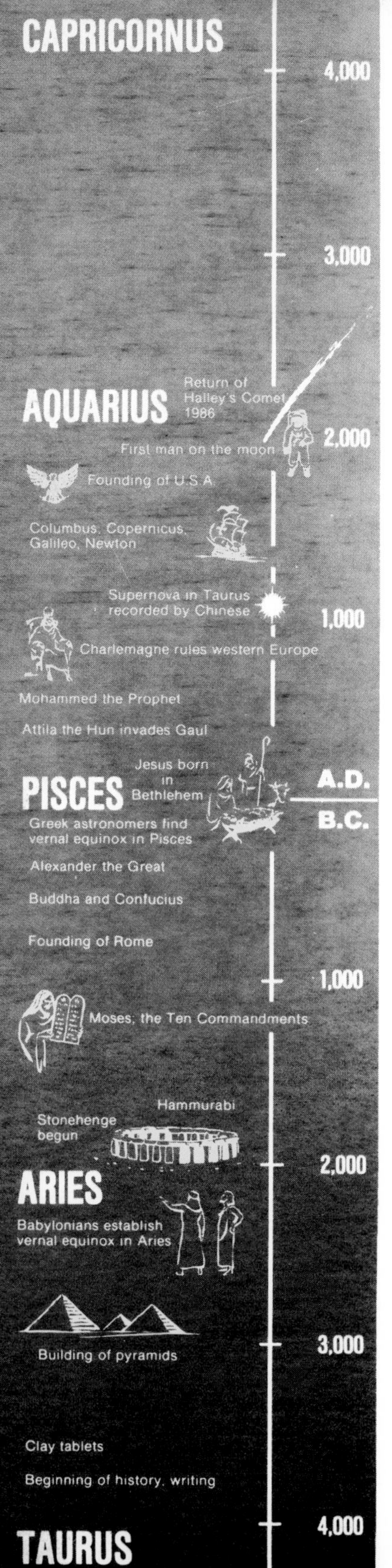


History is measured in man-made centuries. But it might well be measured in star-fixed ages, determined by the natural precessional cycle, with a new Age of Man dawning every 2,200 years as the vernal equinox slips into another constellation. At the beginning of civilization, the vernal equinox was in the Twins; so that period might be called the Age of Gemini. By the time of the Hebrew Patriarch Abraham, the vernal equinox was in the Bull; so this period of history might be called the Age of Taurus. His distant successor, Moses, was a man of another age—the Age of Aries. Moses was angered more than once when, upon descending from his meditations on Mt. Sinai, he found his people turning back to old ways, worshipping the Golden Calf, the "sign of their forefathers"—could this be equated with Taurus, the Bull?

Since the beginning of the Age of Pisces, the sun has crossed the equator about two thousand times to bring the season of spring to the northern hemisphere. Spring has come a little earlier each time, and the vernal equinox has moved gradually to the west side of Pisces, near the Circlet.

The Age of Pisces is drawing to a close. The vernal equinox will be leaving the Fishes for the Water Carrier in a few hundred years. To the astronomer, the event is a matter of position along the ecliptic. To the historian, it represents the fifth of the star-fixed Ages of Man. And to the astrologer, it is the dawning of an age of hope—the Age of Aquarius.

Jupiter and Saturn performed a triple conjunction in Pisces in the year 7 B.C. to open the Age of Pisces. Mars joined them in the spring of 6 B.C. to make a fine triangle low in the western sky near the setting sun. Persian astrologers had long been searching the sky for a sign announcing a predicted great event. Perhaps this conjunction and gathering of planets in the Fishes was the "sign" that started the Wise Men (astrologers) on their journey to King Herod bearing costly gifts.

# KNOWING THE STARS

*Charting constellations,*
*plotting planets,*
*and setting names upon the stars,*
*the ancients*
*accurately mapped the heavens*
*from North Star to Southern Cross.*

*We, of an age*
*less given to myth and legend,*
*see the same random points of light*
*on the celestial sphere.*

*Ideas of old*
*live on in the heavens*
*—place names and interlocking designs,*
*guiding the astronomer*
*to what's beyond.*

The earliest known description of the sky was made early in the 4th century B.C. by the Greek, Eudoxus of Cnidus. His writing has been lost, but his ideas were preserved in verse by the Cilician poet Aratus, in 270 B.C. Dividing the sky into northern, zodiacal, and southern star groups, he described 44 constellations:

### NORTHERN

Ursa Major, Ursa Minor, Draco, Boötes, Cepheus, Cassiopeia,
Andromeda, Perseus, Pegasus, Hercules, Auriga,
Triangulum, Delphinus, Lyra, Cygnus, Aquila, Sagitta,
Corona Borealis, Serpentarius.

### ZODIACAL

Aries, Taurus, Gemini, Cancer, Leo, Virgo, Chelae (now Libra),
Scorpius, Sagittarius, Capricornus, Aquarius, Pisces.

### SOUTHERN

Canis Major, Canis Minor, Orion, Eridanus, Lepus, Centaurus,
Piscis Austrinus, Ara, Cetus, Crater, Corvus, Hydra,
Argo Navis (now Carina, Puppis, and Vela).

A hundred years later, Hipparchus added Lupus,the Wolf to the list, and split Serpentarius into Serpens and Ophiuchus. He also may have added Equuleus and Corona Australis. But he failed to mention Coma Berenices even though it was in existence at the time.

Johann Bayer's *Uranometria* of 1603 showed the positions of 1,250 stars with their relative brightnesses, and 11 new constellations far to the south:

Chamaeleon, Dorado, Hydrus, Piscis Volans (now Volans),
Apis, the Bee (now Musca, the Fly),
Avis Indicia (now Apus, the Bird of Paradise),
Grus, Indus, Phoenix, Tucana, Triangulum Australe.

At about the same time, Jacob Bartsch put Camelopardalis, Monoceros, and Columba in the sky, and reported the existence of Reticulum, the Net. Augustine Royer, detaching that long-recognized cross from Centaurus, invented Crux in 1679.

The Polish astronomer Hevelius published seven new northern groups in 1690:

Canes Venatici, Vulpecula, Scutum, Lacerta,
Lynx, Leo Minor, Sextans.

Nicolas Louis de La Caille added 13 new constellations in 1769, all in the south:

Sculptor, Fornax, Horologium, Caelum, Mensa,
Pyxis, Microscopium, Telescopium,
Pictor, Circinus, Norma, Octans, Antlia.

The sky is now complete. Eighty-eight constellations fill all the available space. That number is not likely to change, for a commission of the International Astronomical Union in 1928 agreed upon their definite though irregular boundaries.

Constellation boundaries of old were loose, drawn with considerable freedom by celestial cartographers. The figure itself, together with its "stars outside," was important in locating objects. Now that has changed. Constellations are precisely defined regions on the celestial sphere (just as states are precisely defined geographic regions on the terrestrial sphere) bounded by arcs of meridians and parallels of declination. A "comet in Capricornus," a "pulsar in Perseus," or a "nova in Cassiopeia," tells in what part of the sky to look. A coordinate system, the celestial equivalent of latitude and longitude, pinpoints the position in terms of right ascension and declination. That information, placed into the setting circles of the telescope, locates the object even before the observatory dome is opened.

## THE CELESTIAL SPHERE

Stars seem to lie on a spherical surface, only half of which we can see at any one time. The *celestial sphere* was real for the ancients, stars being fixed to that shell. We still retain the celestial sphere as a convenient frame of reference for describing locations.

The celestial sphere turns. Stars move across the sky, differing in their rates of motion. Those near the poles rotate slowly in circles; those near the equator move rapidly in great arcs across the heavens. All return to their starting places four minutes earlier each day as the earth circles the sun.

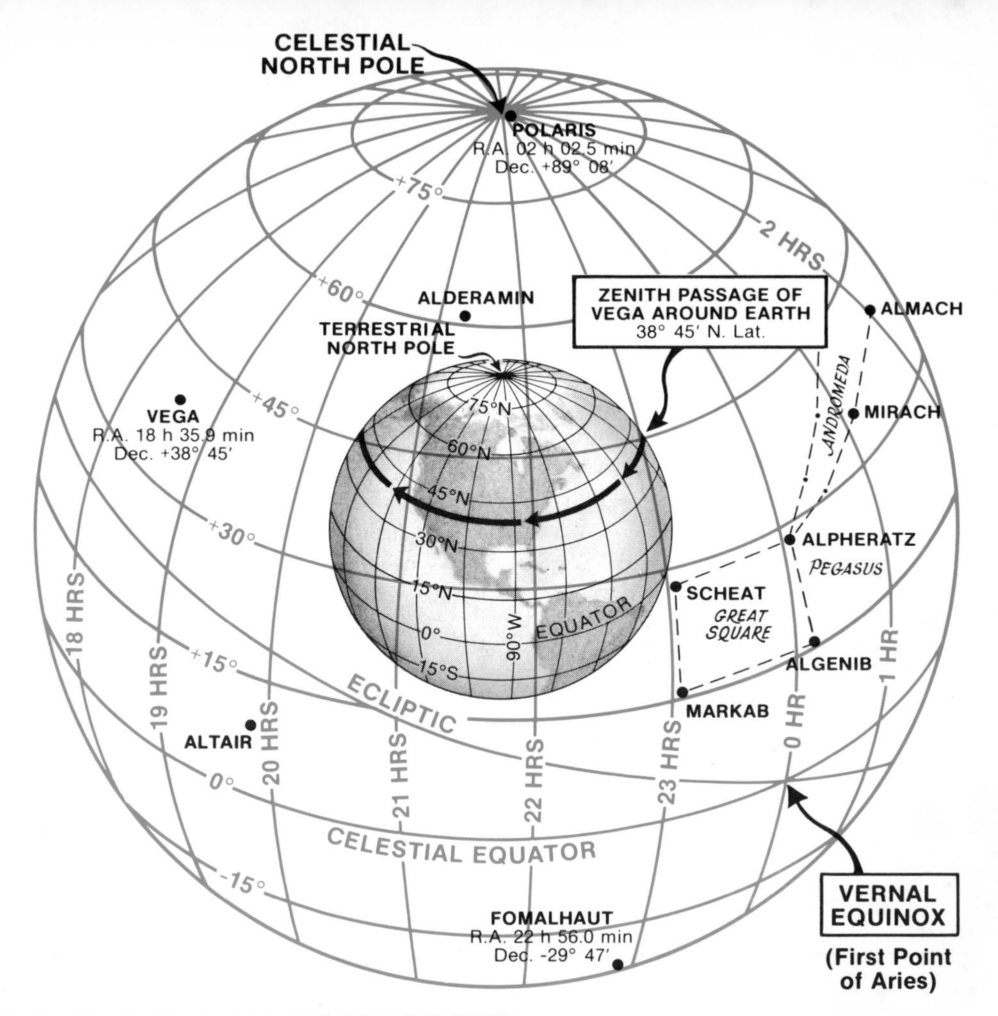


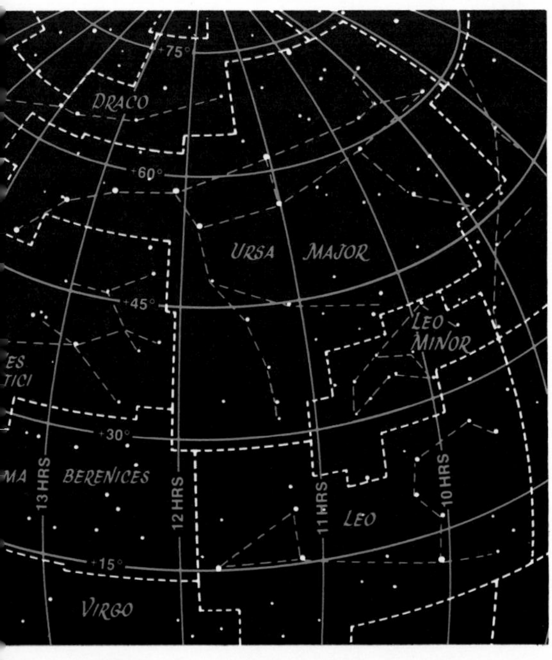


The celestial sphere is divided into 24 *hour circles*—great circles passing through the north and south celestial poles—corresponding to the meridians on earth. Zero hour circle, though, does not pass through Greenwich, but through the First Point of Aries—the vernal equinox. That point gradually shifts westward, and periodically, star positions are refigured.

A star's position is given in *right ascension* and *declination.* Right ascension, measured to the east from the vernal equinox, is expressed in time units. Declination, the angular distance of a heavenly body north or south of the celestial equator, is expressed in degrees.

Left: Observing from earth, we see the celestial sphere from the inside. White dashed lines are constellation boundaries established by international agreement.

## SEASONAL AND MONTHLY MAPS OF THE SKY

Two types of star charts appear in this section: *meridional* and *circumpolar.*

The twelve meridional charts show the sky as it appears along the meridian at 9:00 p.m. in mid-month. They cover the sky from 15 degrees beyond the north celestial pole to within 20 degrees of the south celestial pole. Ninety percent of the celestial sphere is shown—that portion of the sky seen from as far south as the Florida Keys and Hawaii—from North Star to Southern Cross. A special grid has been used to plot the constellations so that they appear with relatively little distortion throughout the months.

The circumpolar charts show the stars within 60 degrees of the pole, as they appear in the four seasons of the year.

### Using the Charts

To find constellations in the night sky, face south and hold the chart of the month overhead. East is to the left, west to the right, and Polaris is behind you to the north. During the night, the stars drift westward across the heavens, from left to right.

The meridional charts picture the sky at two-hour intervals. To find the stars that were overhead two hours ago, look at the previous month's chart. To find those that will be close to the meridian at midnight, look at next month's chart.

### Constellation Configuration

So familiar are celestial figures that have appeared on charts for centuries—Orion, Scorpius, Leo, Taurus—that they are shown in their traditional design-forms; for to do otherwise would seem a sacrilege.

Other constellations appear either in slightly modified image patterns or in entirely new forms. This has been done by (1) using more of the stars within that area of the sky allotted to each constellation by the 1928 International Agreement, and (2) reshaping the image to make it more closely resemble the idea that it represents, such as Antlia, Microscopium, Phoenix, and Telescopium.

Curving across each chart is the ecliptic, the sun's annual path through the stars. Equinoctial and solstitial positions are shown as they are in the 20th century. The celestial equator, though not shown, can be inferred from the solstitial and equinoctial points.

The Milky Way, tilted 62 degrees to the celestial equator, is shown as a faint band of blue stars.

## Messier Objects

Two hundred years ago, comet-watcher Charles Messier cataloged 103 bright, fuzzy objects in the sky so that he wouldn't confuse them with comets. His system, since increased by six, is still in use today even though other systems have been introduced. Messier objects are designated M1, M2 . . . M109. We now recognize them as star clusters, nebulae, and galaxies. Only the brightest and most important Messier objects are shown on the charts—those that can be seen with the unaided eye, binoculars, or small telescopes up to 3-inch.

## Stellar Magnitudes

Principal stars down to 6th magnitude are shown in white on the charts. Fainter, random stars are shown in blue. The unaided eye can see stars of 6th magnitude under the best of viewing conditions. With binoculars, you can see stars of 8th or 9th magnitude. A 6-inch telescope reveals stars of 13th magnitude.

Six thousand stars of 6th magnitude or brighter are visible to the unaided eye. Since only half the celestial sphere can be seen at any time, that number is cut to 3,000. Many stars are near the horizon, and are hard to see in the thickness of the atmosphere, reducing the number to perhaps 2,500. But add a full moon and city lights, and the number decreases to a few hundred. Stars below 4th magnitude (the magnitude of Alcor, companion of Mizar in the Big Dipper's handle) are seldom seen when viewing from cities.

A difference in magnitude of 1 is equivalent to a brightness difference of 2.5. The magnitudes of stars shown in the charts follow this scale:

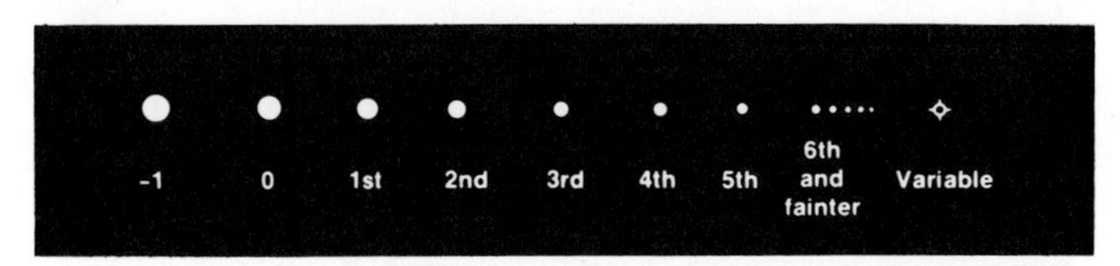


Stars are so far away that they're only bright pinpoints of light even in the biggest telescopes on earth. And they twinkle—flashing red, blue, green, yellow as starlight is bent one way and another in traveling through the cells of hot and cold air in the earth's atmosphere. Planets do not twinkle. They have observable diameters and glow steadily, brilliantly.

# stars of WINTER

Winter positions of stars near the North Star.

High in the heavens on winter evenings are eight of the 13 most brilliant stars, and many of the best-known constellations.

Orion dominates the scene. Standing at the edge of the River Eridanus, the Giant battles Taurus, the Bull. Behind him are his two dogs. A timid rabbit and a dove are beneath him, and above his head are the Twins. Two other heroic figures are nearby—Perseus rescuing Andromeda, and the compassionate chariot-driver Auriga.

A colorful array of 1st magnitude stars gathers about Orion—brilliant blue-white Rigel, warm-orange Betelgeuse, whitish-green Castor, sun-yellow Pollux and Capella, red-orange Aldebaran, and pale-yellow Procyon.

A line through the belt of Orion extended westward reaches the Bull's eye; and in the opposite direction, the brightest star in the sky, Sirius. The stars in Orion's belt are the Three Canoe Paddlers in Polynesia; here is the only place in the sky where three evenly-spaced bright stars lie in a row.

Each star in Orion's belt has its individuality. Mintaka, the westernmost, lies on the celestial equator, dividing the Giant between the two hemispheres. The center star, Alnilam, is one of the hottest stars known—about 45,000° Kelvin. And from the third star, Alnitak, hangs the sword of Orion.

The fuzzy patch of light in Orion's sword is the Great Nebula—a pale-green mass of interstellar gas and dust illuminated by four bright stars within—a beautiful sight in a telescope. The four very blue stars in a tiny square, the *Trapezium,* are sending out high-energy particles that hit the gaseous cloud, exciting it into brilliance. The Great Nebula shines as a fluorescent object—absorbing energy from stars embedded within it and re-radiating it as light. The dusty nebula in the nearby Pleiades, on the other hand, shines by reflecting light.

# JANUARY

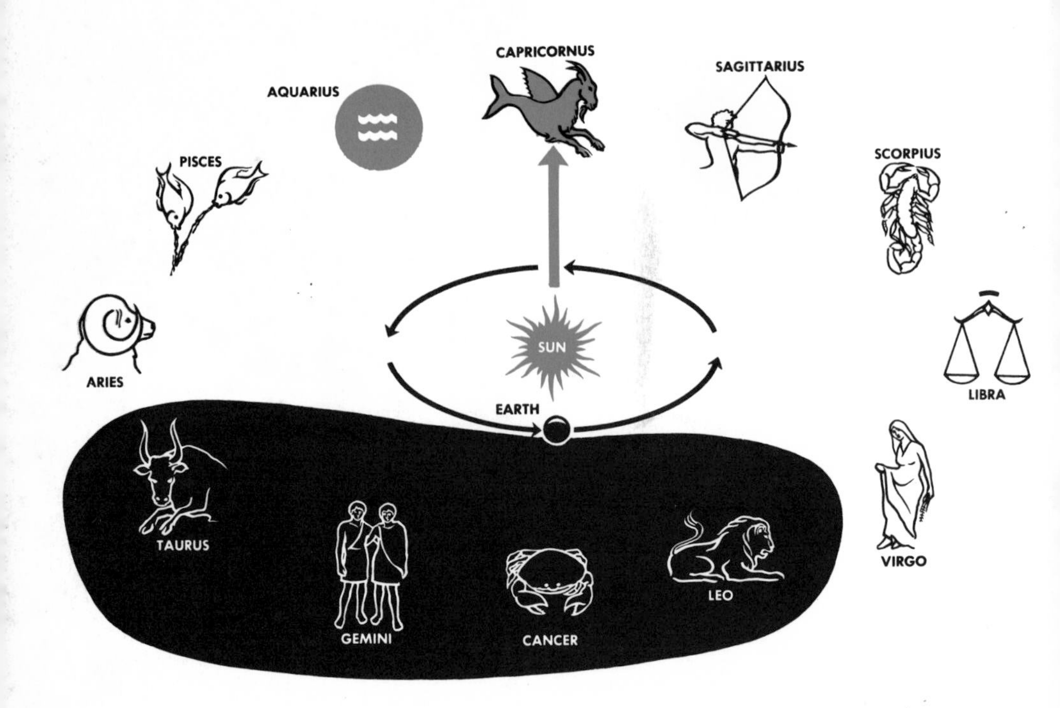


The sun is in the direction of Capricornus
from January 18 to February 14.
Aquarius is the astrological sign
from January 20 to February 18.

Taurus is in the evening sky
along with Auriga and Eridanus.

Giraffe and Graving Tool are near the meridian
early in the evenings at mid-month.

M 37 and M 38 in Auriga, the Pleiades and Hyades
clusters, and the Crab Nebula in Taurus
are objects of interest
in binoculars or telescopes.

**TAURUS, THE BULL,** head lowered and eye ablaze, is poised to lunge at Orion. The Bull was worshipped by Babylonian astrologers. The vernal equinox was in Taurus at the time of the building of the Pyramids.

The V-shaped face of the Bull is a cluster of red and yellow giant stars—the Hyades—140 light-years distant.

Another cluster is the Pleiades, or Seven Sisters. Daughters of World-shouldering Atlas, the Sisters were the objects of Orion's affection. To discourage Orion, Atlas put Taurus in the way. Giant and Bull have now been encountering each other for thousands of years without apparently having done battle.

Most observers see only six stars in the Pleiades. Binoculars reveal fifty. Large telescopes photograph an open cluster of 3,000 stars wrapped in faint, luminous dust clouds, traveling together through space like a flock of birds. Stars are being born in this dusty region, 430 light-years from us.

Aldebaran
HYADES
PLEIADES

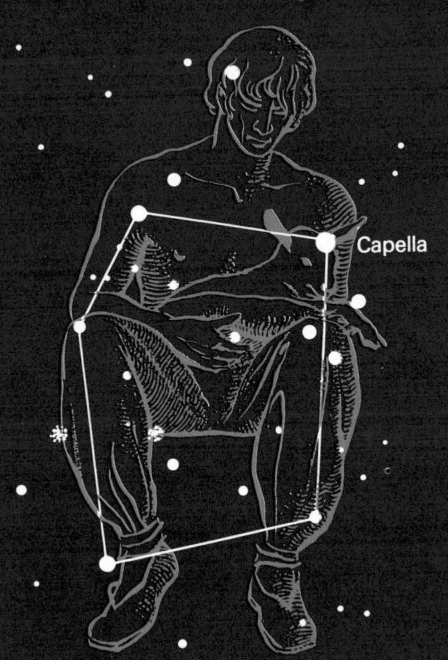


**AURIGA, THE CHARIOTEER** spans the Milky Way—a big, five-sided fellow who reminds us of the qualities of inventiveness and compassion.

Lame as a lad and unable to get around very well, Auriga used his talent to invent the chariot. A lover of animals and people, he holds the young goat, Capella, tenderly in his arms. Stars near Capella are "The Kids."

Capella appears as a single point of light. Actually, it's a pair of yellow giants the color of the sun—ten times larger, fifty times brighter, and three million times farther.

**ERIDANUS, THE RIVER** is a long, meandering stream of stars, ending in bright Achernar—so far to the south that it can be seen only from the southern states. Eridanus (poetic name for the Po River) was the river into which young Phaëthon fell in his unsuccessful attempt to drive the sun across the sky.

The River Eridanus meanders back and forth across the meridian. Sharp-pointed Perseus leads Orion and his brilliant entourage, and the Bull backs his way across the sky.

Jupiter took the form of a snow-white bull to capture his love, Europa. Entranced with the friendliness of the bull, she petted him fondly and seated herself upon his back, whereupon he suddenly flew away with her across the sea to Crete where he changed back to his real form and she became his wife.

Five thousand years ago, Aldebaran, the eye of the Bull, marked the vernal equinox; it was one of the Four Royal Stars of ancient astrology, along with the heart of Leo (Regulus), the heart of Scorpius (Antares), and the eye of the Southern Fish (Fomalhaut). Generally, Taurus was considered an unfortunate sign in astrology, but the almanac of 1386 states that "whoſo if born in yat syne shal have grace in bestis." And if thunder occurred when the sun was in Taurus, then there would be "a plentiful supply of victuals."

CAMELOPARDALIS, THE GIRAFFE is near the north pole, far from his terrestrial habitat in southern climes. It covers a large part of the sky, but its brightest star is only 4th magnitude.

CAELUM, THE GRAVING TOOL was formed by La Caille from the stars between Dove and Clock. It culminates on the 10th of the month, far south of Aldebaran. Its extreme southern position and faintness make it difficult to see for most northern observers.

Lying within the pentagon of Auriga is M38, an open cluster just visible to the unaided eye, at a distance of 3,600 light-years. At Auriga's side is M37, a brilliant open cluster 4,200 light-years distant, excellent in small telescopes.

Capella, the Goat Star in Auriga, is the she-goat of mythology who suckled the infant Jupiter. One day in play, the child broke one of the goat's horns. Immediately he endowed it with the power of being filled with whatever anyone might wish, and thus it became the Horn of Plenty—Cornucopia.

Running between the horns of Taurus is the ecliptic; the Crab Nebula M 1 is just about on it. M 1 is the remnant of a spectacular supernova explosion recorded by Chinese astronomers in A.D. 1054 Within the nebula, 4,000 light-years distant, is a pulsar sending out pulses of radiation thirty times a second. Optical verification of the pulsar has recently been established—a neutron star about 10 km in diameter and a million-million times as dense as water.

Phecda
Megrez
BIG DIPPER
Merak
Dubhe
URSA MAJOR
M81
DRACO
Kochab
LITTLE DIPPER
URSA MINOR
Polaris
DRACO
CYGNUS
Deneb
Alderamin
CEPHEUS
M39
Delta Cephei
VARIABLE
LACERTA
CAMELOPARDALIS
CASSIOPEIA
Caph
Navi
Schedar
M 103
Ruchbah
LYNX
DOUBLE CLUSTER
PERSEUS
M31
ANDROMEDA
Scheat
Capella
Mirfak
Almach
GREAT SQUARE
M44
BEEHIVE
Castor
KIDS
Pollux
AURIGA
Algol
VARIABLE
M34
Alpheratz
Mirach
Markab
Ecliptic
M37
M38
GEMINI
TRIANGULUM
M33
PEGASUS
Elnath
M35
M1
ARIES
Algenib
June Solstice
TAURUS
M45
PLEIADES
Hamal
Sheratan
PISCES
Aldebaran
CANIS MINOR
ORION
HYADES
Ecliptic
Betelgeuse
Bellatrix
Menkar
March Equinox
CEROS
Alnilam
Alnitak
Mintaka
Mira
VARIABLE
M42
CETUS
Saiph
Rigel
Diphda
Sirius
Mirzam
LEPUS
M41
CANIS MAJOR
ERIDANUS
FORNAX
Adhara
Phact
SCULPTOR
COLUMBA
Acamar
CAELUM
Ankaa
PHOENIX
HOROLOGIUM
Canopus
PICTOR
DORADO
Achernar
TUCANA
RETICULUM
Large Magellanic Cloud
HYDRUS
January

# FEBRUARY

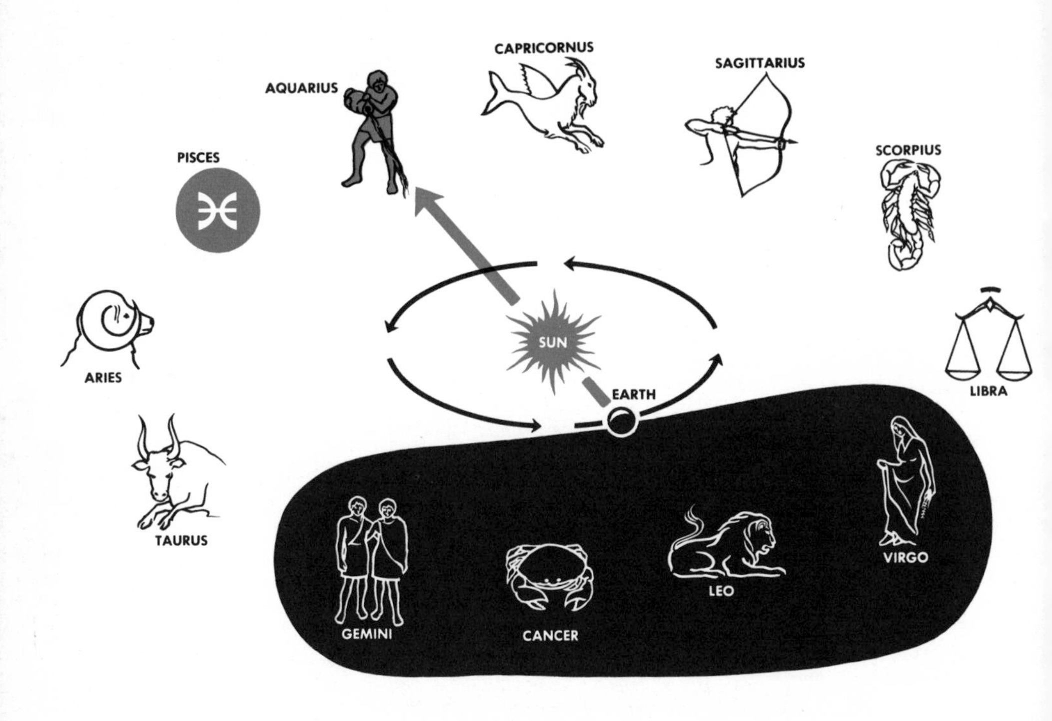


The sun is in the direction of Aquarius
from February 14 to March 14.
Pisces is the astrological sign
from February 19 to March 20.

Gemini is in the evening sky
along with Orion and Lepus.

Dove, Net, Goldfish, and Easel
are near the meridian
early in the evenings at mid-month.

M 35 at the foot of Castor, M 41 in Canis Major,
and the Great Nebula in Orion
are interesting objects.

**GEMINI, THE TWINS**—two bright stars marking their heads, parallel strings of fainter stars, their bodies—are mirror-images of each other, suggesting devotion and interdependence.

Pollux, the brighter of the two, is 35 light-years distant. But Castor, his duller brother, is the more interesting of the pair. Half again as far into space, Castor is a triple star with each of its components double—six stars in all!

The sun at summer solstice (June 22) is at the foot of Castor.

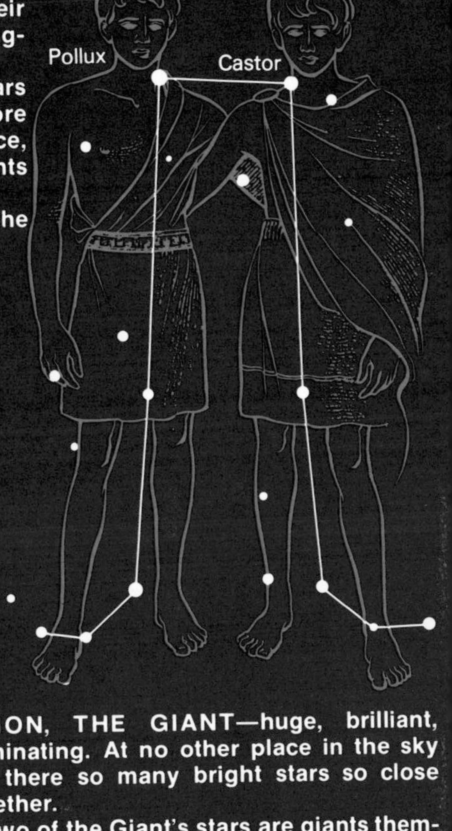


**ORION, THE GIANT**—huge, brilliant, dominating. At no other place in the sky are there so many bright stars so close together.

Two of the Giant's stars are giants themselves. Betelgeuse, brighter than ten thousand suns, is a red giant that shrinks and swells from 700 to 1,000 sun diameters. Now old, Betelgeuse will soon puff off its outer layer of hydrogen, then collapse into a white dwarf star. Our sun will do the same in another five to ten billion years.

Rigel, a giant of a different sort, is burning up its fuel at a reckless rate. Extremely hot (25,000° C) and brighter than 60,000 suns, Rigel will eventually become unstable and explode. Total lifetime: perhaps ten million years—a very short life for a star.

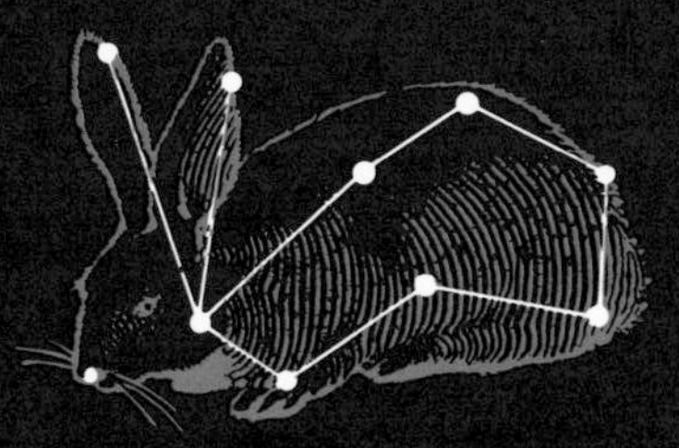

**LEPUS, THE HARE** rests quietly at Orion's feet. All but lost in the brilliance of the Giant's stars, he seems unconcerned with the antics of either hunter or dogs.

A line from Polaris crosses the ecliptic at the summer solstice and runs southward through Betelgeuse and Canopus. The faintest part of the Milky Way is in the sky, and the two brightest stars, Sirius and Canopus, reach the meridian.

Castor and Pollux were the sons of Jupiter and Leda, and brothers of Helen of Troy. Castor was the manager of horses; Pollux, a famous soldier. The centaur Chiron trained them, and they sailed with the Argonauts in search of the Golden Fleece. During a storm on the voyage when the ship was in great danger, the two stars flamed forth on the heads of the Twins—a heavenly sign that all was well. Ever since then the sky couple has been associated with electrical phenomena that appear in storms and heavy seas—a double light to sailors. The flame-like discharge from the masts and spars of ships is known today as St. Elmo's Fire.

Astrologers consider the sign of the Twins to be favorable, but 500 years ago a person born under it would be "ryght pore and wayke and lyf in mykul tribulacion."

COLUMBA, THE DOVE was formed by Royer in 1679 near *Argo,* the ship. Originally it was *Columba Noae*—Noah's Dove—and was supposed to be the dove that Noah sent out from the ark.

Close to the position of the summer solstice is M 35, an open cluster of 6th magnitude stars at a distance of 2,500 light-years. Triangular in small telescopes and diamond-shaped in large, M 35 is a good object in field glasses.

Two of the three planets discovered with the telescope were found in Gemini. Sir William Herschel found Uranus in 1781 near M 35. Pluto was predicted, and after years of search, was finally photographed at the foot of Castor in 1930 by the staff of the Lowell Observatory.

RETICULUM, THE NET lies northwest of the Large Magellanic Cloud. Nearby is DORADO, THE GOLDFISH—not our "little exotic cyprinoid, but the large coryphaena of the tropical seas, changing colors on death." It has also been called the Swordfish. Within it is the south ecliptic pole. PICTOR, THE EASEL is a La Caille constellation between Canopus and the Large Cloud of Magellan.

M 41 in Canis Major can be seen in a small telescope; it is an open cluster of fifty 8th magnitude stars.

The Great Nebula in Orion (M 42) is a quadruple star system in a diffuse nebulosity, a thousand light-years distant.

Alkaid
LITTLE DIPPER
Alcor
Thuban
Kochab
CEPHEUS
LACERTA
M 51
Mizar
DRACO
Delta Cephei
VARIABLE
Alioth
URSA MINOR
CANES VENATICI
Megrez
Caph
Polaris
Phecda
BIG DIPPER
Navi
Schedar
M 81
M 103
CASSIOPEIA
Merak
Dubhe
CAMELOPARDALIS
Ruchbah
M 31
Alpheratz
URSA MAJOR
DOUBLE CLUSTER
ANDROMEDA
Mirach
PERSEUS
Mirfak
M 34
Almach
M 33
LEO MINOR
LYNX
Capella
Algol
VARIABLE
PISCES
KIDS
TRIANGULUM
AURIGA
SICKLE
Hamal
Sheratan
Algenubi
M 37
M 38
Castor
ARIES
M 45
PLEIADES
Pollux
GEMINI
Elnath
CANCER
M 35
Ecliptic
M 44
BEEHIVE
TAURUS
Regulus
Ecliptic
M 1
June Solstice
Aldebaran
HYADES
Menkar
ORION
Bellatrix
CETUS
Procyon
CANIS MINOR
Berelgeuse
Mira
VARIABLE
HYDRA
Alnilam
Mintaka
Alnitak
M 42
MONOCEROS
Rigel
Alphard
Saiph
ERIDANUS
Sirius
Mirzam
LEPUS
M 41
CANIS MAJOR
Adhara
FORNAX
PYXIS
Phact
COLUMBA
PUPPIS
CAELUM
Acamar
Suhail
VELA
Canopus
PICTOR
HOROLOGIUM
DORADO
RETICULUM
CARINA
Avior
HYDRUS
Large Magellanic Cloud
February

# MARCH

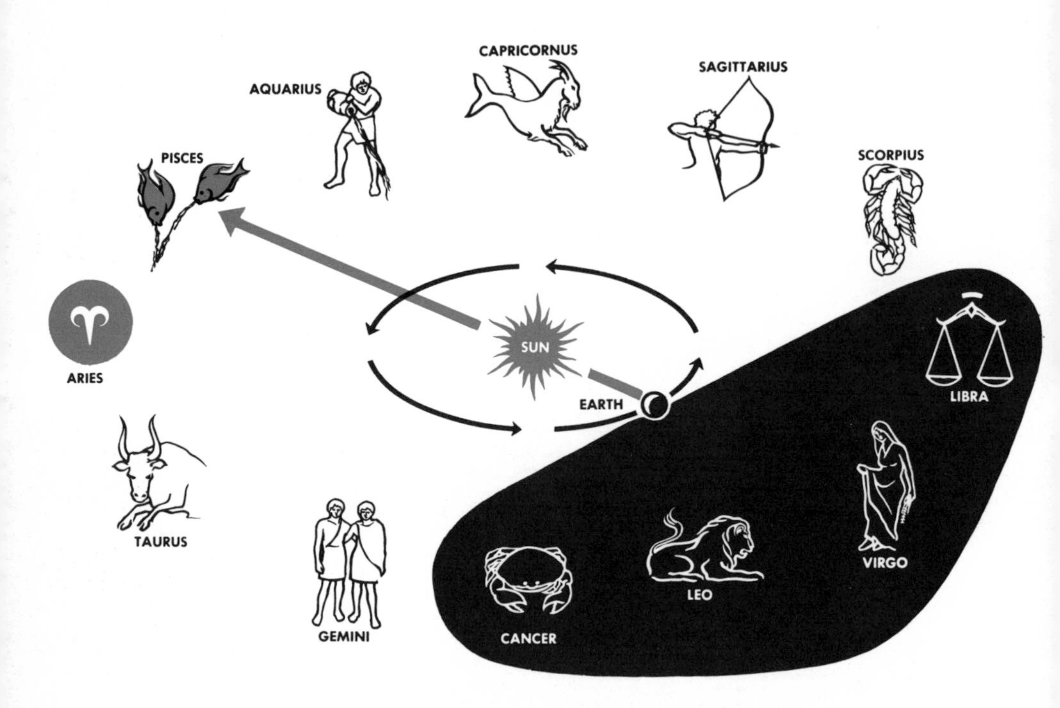


The sun is in the direction of Pisces
from March 14 to April 15.
Aries is the astrological sign
from March 21 to April 19.

Cancer is in the evening sky
along with Canis Major and Canis Minor.

Lynx, Unicorn, Stern, Sail, Compass, and Keel
are near the meridian
early in the evenings at mid-month.

The Praesepe Cluster, M 44,
is an interesting object in binoculars.

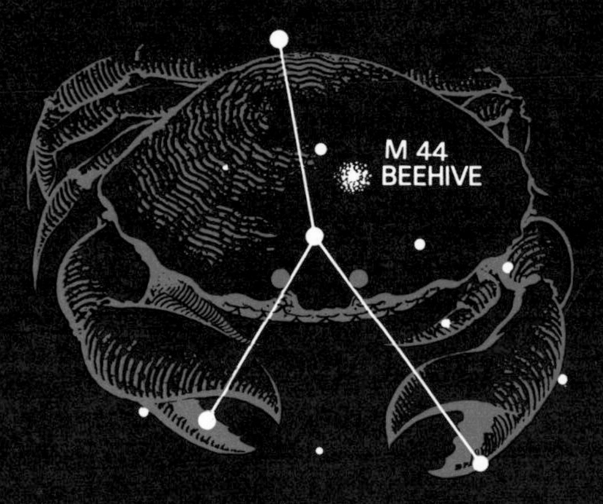


**CANCER, THE CRAB** is a faint group shaped like an upside-down Y. It is hard to find, but well worth finding for its Praesepe Cluster, commonly known as the Beehive.

The Beehive is an open cluster of more than 400 very faint stars at a distance of 500 light-years. Its faintness is sometimes used in predicting weather, for even a slight collection of water vapor in the air will obscure it.

The sun in ancient times was in the direction of Cancer at summer solstice. Now it is in Gemini.

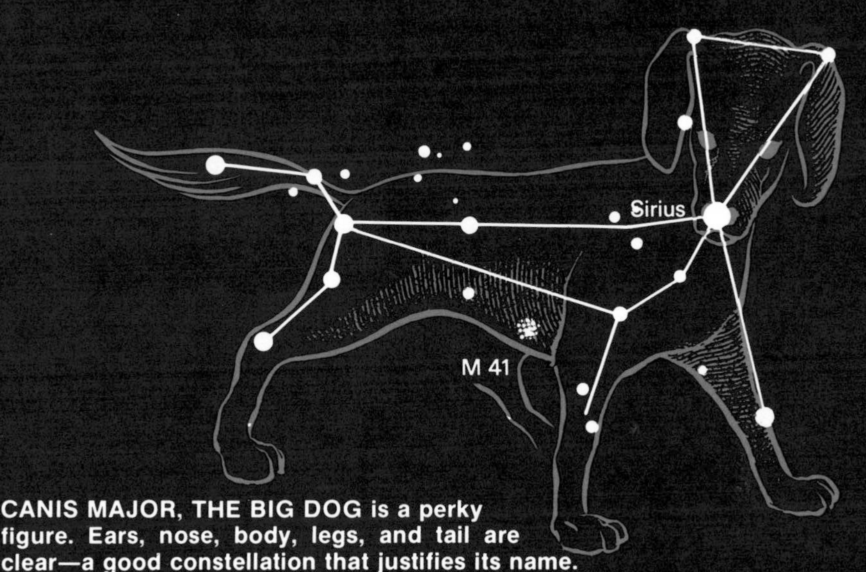


**CANIS MAJOR, THE BIG DOG** is a perky figure. Ears, nose, body, legs, and tail are clear—a good constellation that justifies its name.

Blue-white Sirius is the brightest star in our sky. Much of its brilliance is due to the fact that it is so close to us, only 8.7 light-years distant. It is also a bright star in its own right—a star twice the sun's diameter and three times as hot.

Companion to the "Dog Star," Sirius, is a white dwarf star, "The Pup." The unseen companion is a star of such great density (a teaspoonful weighs a ton!) that it causes the Dog Star to wobble through the sky—a celestial example of a tail wagging the dog.

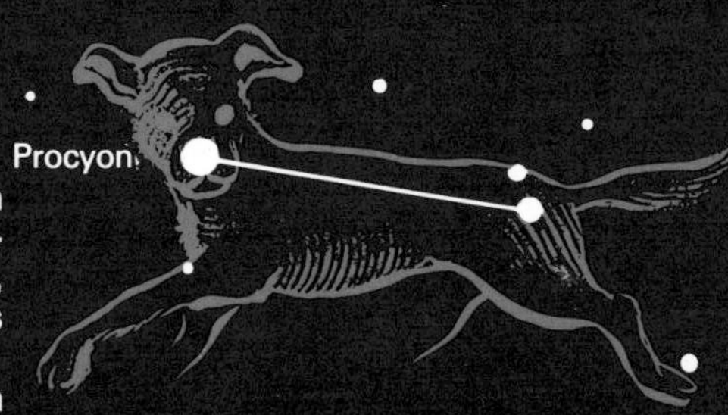


**CANIS MINOR, THE LITTLE DOG** rides on the back of Monoceros, the Unicorn. First-magnitude Procyon, 11 light-years distant, and a faint group of stars—that's all there is to this little constellation.

Procyon, Sirius, and Betelgeuse make a bright equilateral triangle in the sky.

Riding the head of Hydra is the toe-pinching Crab who helped the Sea Serpent battle Hercules by nipping the Hero's feet—eventually to be crushed by the giant and put into the sky as a faint group of stars.

Cancer was the Gate of Men through which the souls of mortals descended from heaven into human bodies. But it was an unfortunate sign in astrology, governing the human breast and stomach. The sun in Cancer meant that thunderstorms would cause commotions, famine, and locusts. One observer feared that the earth would be submerged when all the planets gathered in Cancer.

Between the Twins and Bear, Lion and Giraffe is a dark region with only a few faint stars. Hevelius formed LYNX out of the stars there, explaining that the observer had to be lynx-eyed in order to see it. Many beautiful doubles lie within the constellation.

MONOCEROS, THE UNICORN rides the back of the Big Dog. The group is not conspicuous, but within it is the Rosette Nebula—a shell of red hydrogen gas surrounding a brilliant white star. The nebula is 4,000 light-years distant, and a photographic object in large telescopes.

ARGO NAVIS THE SHIP was once a huge constellation far to the south, backing across the sky, its stern beneath the hindquarters of Canis Major. Later it was split into three smaller, more manageable groups—PUPPIS, THE STERN; VELA, THE SAIL; and CARINA, THE KEEL. Argos built the ship for young Jason to help him in his quest for the Golden Fleece. Jason gathered up fifty men—Hercules, Cepheus, Orpheus, Castor, and Pollux among them—and set out on a series of exploits that finally got them the fleece of Aries the Ram. Upon their return, the Goddess Athena placed Argo Navis in the sky, dedicating it to Neptune.

The Keel, Stern, and Sail are all that remain of the old ship. Its bow may have been crushed as it dashed between the Clashing Islands of mythology. *Pyxis Nautica,* THE MARINER'S COMPASS (now PYXIS) is sometimes associated with Argo Navis. But as such, it is a celestial anachronism, for the compass was not invented until long after the time of the Argonautic expeditions.

Brilliant Canopus in the Keel (the second brightest star in the sky) leads the procession. The star was named after Canopus, chief pilot in Menelaus' fleet which destroyed Troy in 1184 B.C. Canopus culminates 20 minutes before Sirius, but it is so far south of the celestial equator (53 degrees) that the observer in the southern states sees it only briefly.

LITTLE DIPPER
CEPHEUS
Caph
M31
Schedar
Navi
CASSIOPEIA
Ruchbah
Kochab
URSA MINOR
Thuban
DRACO
Polaris
ANDROMEDA
Mirach
M103
Alkaid
Alcor
Mizar
DOUBLE CLUSTER
M51
CAMELOPARDALIS
Almach
Alioth
M81
CANES VENATICI
Megrez
TRIANGULUM
Caroli
Phecda
BIG DIPPER
Dubhe
Mirfak
M34
Merak
Algol
VARIABLE
URSA MAJOR
ARIES
Capella
PERSEUS
Hamal
KIDS
AURIGA
LYNX
M45
PLEIADES
M37
M38
Ecliptic
LEO MINOR
Elnath
Castor
M1
M35
TAURUS
Pollux
GEMINI
June Solstice
Aldebaran
HYADES
SICKLE
Algenubi
LEO
CANCER
M44
BEEHIVE
Bellatrix
Regulus
ORION
Betelgeuse
Procyon
CANIS MINOR
Alnilam
Mintaka
Alnitak
M42
ERIDANUS
SEXTANS
Rigel
MONOCEROS
Alphard
Saiph
HYDRA
Sirius
Mirzam
LEPUS
M41
ANTLIA
CANIS MAJOR
PYXIS
Adhara
Phact
PUPPIS
COLUMBA
Suhail
CAELUM
VELA
Canopus
PICTOR
Avior
CARINA
DORADO
VOLANS
Large Magellanic Cloud
March

# stars of SPRING

Spring positions of stars near the North Star.

The brilliant stars of winter slide farther west each evening as spring approaches. Canopus sets, and Sirius twinkles brightly for a while before disappearing into the twilight as the sun moves eastward along the ecliptic. Orion looks even more imposing as he nears the horizon; but he, too, fades before the advancing sun.

We turn our backs on the winter stars as they're setting in the west to see the great stars that will rule the summer sky, rising in the east. Between these two spectacular stellar shows are the stars of spring—not brilliant, but containing the two most famous star groups: the Big Dipper and the Southern Cross.

A great, gentle curve unites the two. The Big Dipper's handle forms an arc that sweeps through Arcturus and Spica, then continues far down into the southern hemisphere to the Southern Cross, only 30 degrees from the south celestial pole.

Each winter evening, the Big Dipper swings farther southward; and in the spring, its two pointers reach the meridian. The Big Dipper is definitely a star group of the north, with all its stars passing over Canada. Alkaid, the end star in the Dipper's handle, travels along the western border of the United States and Canada.

The names of the stars in the Big Dipper are Arabic in origin. Recite them—*"Alkaid, Mizar, Alioth, Megrez, Phecda, Merak,* and *Dubhe"*—and in the rhythm of the cabalistic chant it seems as if the heavens might open up, or a genie appear.

The Lion lies beneath the Great Bear (The Big Dipper is only the brightest part of Ursa Major). Regulus, "The Heart of Leo," signifies the season as one of the four Royal Stars of ancient astrology.

Denebola (the end star in Leo's tail), along with Arcturus and Spica, form a huge triangle in the east that replaces the brilliant Sirius-Procyon-Betelgeuse equilateral now setting. Add Cor Caroli to the Denebola-Arcturus-Spica spring trio to form a diamond-shaped asterism, the *Diamond of Virgo,* 50 degrees in length.

Omega Centauri, "a noble globular cluster; beyond all comparison the richest and largest in the heavens," according to Herschel, contains thousands of rose-colored stars, 20,000 light-years distant. However, the cluster lies too far south to be seen from mid-latitudes.

# APRIL

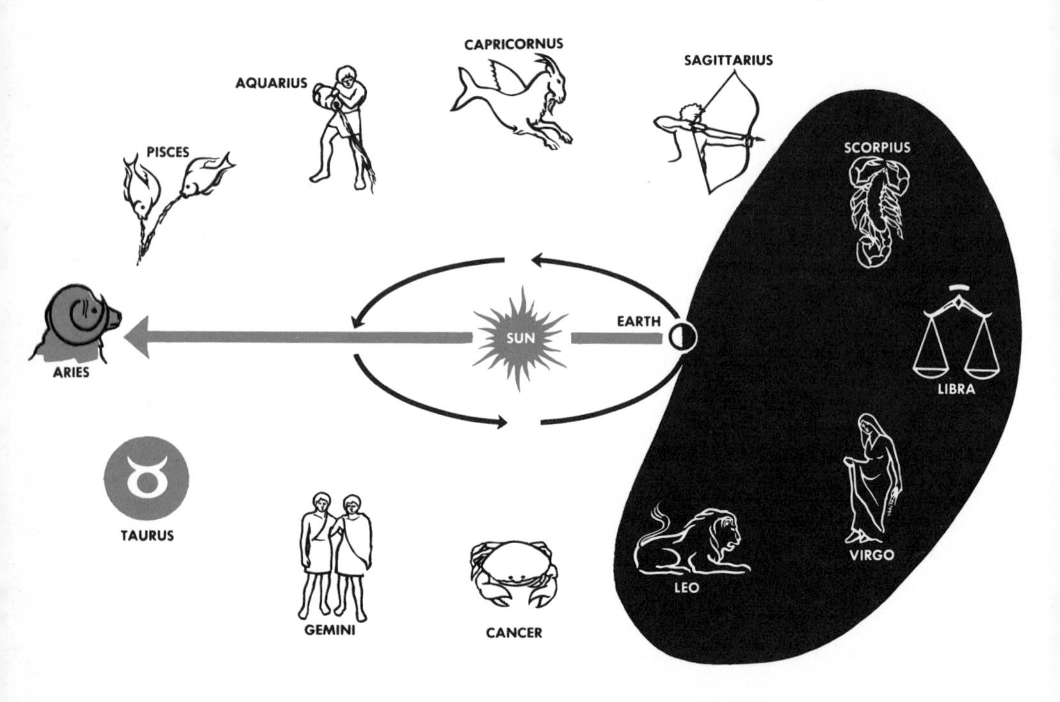


The sun is in the direction of Aries
from April 15 to May 13.
Taurus is the astrological sign
from April 20 to May 20.

Leo is in the evening sky
along with Ursa Major and Hydra.

Little Lion, Sextant, Air Pump, and Flying Fish
are near the meridian
early in the evenings at mid-month.

M 81 is a tightly-coiled spiral galaxy
in the northern sky.

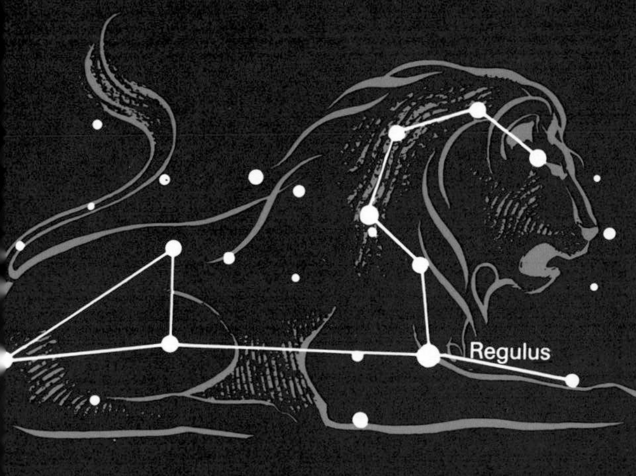


**LEO THE LION**, lying right beneath the Big Dipper, is a large, bright star group along the sun's path.

Curving like a backward question mark in the sky is a group of stars forming the head and mane of the Lion. Blue-white Regulus, 84 light-years distant, is the Heart of Leo, *Cor Leonis.*

The Leonid meteor shower radiates from the center of The Sickle in mid-November as the earth annually runs into debris left by a passing comet.

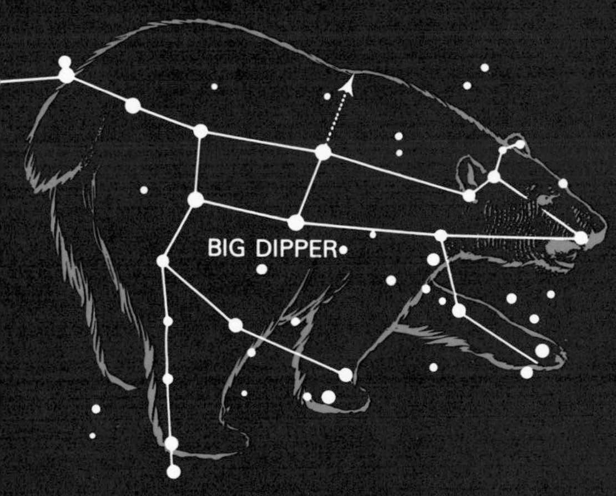


**URSA MAJOR, THE GREAT BEAR** lies above Leo, so far to the north that all its stars pass over Canada. The word *arctic* comes from the Greek word for bear.

The seven brightest stars in Ursa Major form the best-known star group. But it's not a constellation! The Big Dipper is an *asterism*—a distinctive group, but not one of the 88 recognized constellations.

All but the two end stars of the Big Dipper are members of a gigantic moving cluster. The sun lies within the enormous group but is not actually a member of it.

**HYDRA, THE SEA SERPENT** is a straggling string of faint stars wandering a quarter of the way across the sky. Orange-red Alphard, the Solitary One, is its only bright star.

Hercules slew the nine-headed serpent that grew two heads any time one was lopped off. Hydra signifies a persistent evil with many sources.

Above Hydra's back are two groups: Crater and Corvus. Crater is the Cup of Bacchus carried by the Cup-Bearer, Ganymede. Corvus, the Crow—a lopsided square of stars—is the bird that served as a messenger to Apollo.

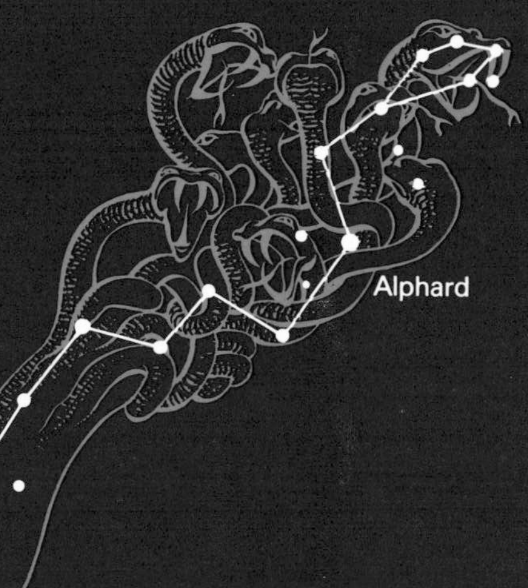

Leo is an ancient constellation that was identified with the sun, "because the sunne being in that signe is most raging and hot like a lion." The sun is now in the direction of Leo in late August. Physicians of old thought that medicine was a poison and a bath harmful when the sun was in the sign. Leo represents the Nemean Lion which Hercules slew in one of his famous labors. And the Great Sphinx at Giza, some say, is Virgo's head on Leo's body.

Most stars have Arabic names, but the brightest one in Leo is an exception. Copernicus named it Regulus, "Little King," in the belief that it ruled the heavens as one of the Four Royal Stars. "The Lyon's herte is called of some men, 'the Royall Starre,' for they that are borne under it, are thought to have a royall nativitie." It was the study of this star that revealed to Hipparchus the precession of the equinoxes.

Above Leo's head and beneath the hind paws of the Big Bear is LEO MINOR, THE LITTLE LION. Farther north in Ursa Major is a spiral galaxy with a glowing nucleus, M 81, some 6,500,000 light-years distant.

Conquering the multi-headed Hydra (not Hydrus) was another of Hercules' great labors. Whenever one of its heads was cut off, Hydra immediately grew two in its place—severely escalating the problems of the would-be conquerer. Iolaus, giving him an assist, quickly seared the spot with a hot iron each time Hercules severed a head, preventing another from growing. Cor Hydrae (Hydra's Heart) is 2nd magnitude Alphard, a bright star 94 light-years distant.

SEXTANS, THE SEXTANT rides the back of Hydra. Hevelius named it, immortalizing an instrument that he used in his stellar measurements between 1658 and 1679. But, "with more zeal than taste, he fixed the machine upon the Serpent's back," placing it between Hydra and the fiery astrological sign of Leo. Hevelius lost his astronomical instruments when "Vulcan overcame Urania" in 1679 and his house was destroyed by fire. His sextant, though, is now preserved in the celestial vault.

*Antlia Pneumatica* (now ANTLIA, THE AIR PUMP) was named *Machine Pneumatique* by La Caille. It is an inconspicuous group that has no named stars. Neither has *Piscis Volans* (now VOLANS, THE FLYING FISH); its brightest star is 4th magnitude.

M 103
CASSIOPEIA
DOUBLE CLUSTER
M 34
Algol
VARIABLE
DRACO
LITTLE DIPPER
Polaris
PERSEUS
Mirfak
URSA MINOR
Kochab
CAMELOPARDALIS
Thuban
M 81
Capella
KIDS
AURIGA
Alcor
Alkaid
Mizar
Alioth
Megrez
Dubhe
BIG DIPPER
BOÖTES
M51
Phecda
Merak
M 38
URSA MAJOR
TAURUS
CANES VENATICI
Elnath
M 37
LYNX
Cor Caroli
M 1
M3
M 35
June Solstice
Arcturus
Castor
GEMINI
COMA BERENICES
Pollux
Ecliptic
LEO MINOR
M 44
BEEHIVE
Algenubi
SICKLE
CANCER
LEO
Betelgeuse
ORION
Denebola
Regulus
Realm of Galaxies
CANIS MINOR
Procyon
VIRGO
September Equinox
Ecliptic
SEXTANS
MONOCEROS
Spica
Alphard
Sirius
CRATER
Mirzam
Gienah
CANIS MAJOR
CORVUS
HYDRA
M41
ANTLIA
PYXIS
Adhara
Suhail
PUPPIS
VELA
CENTAURUS
OMEGA CENTAURI
SOUTHERN CROSS
Canopus
Gacrux
Avior
CARINA
PICTOR
COAL SACK
Acrux
VOLANS
April

# MAY

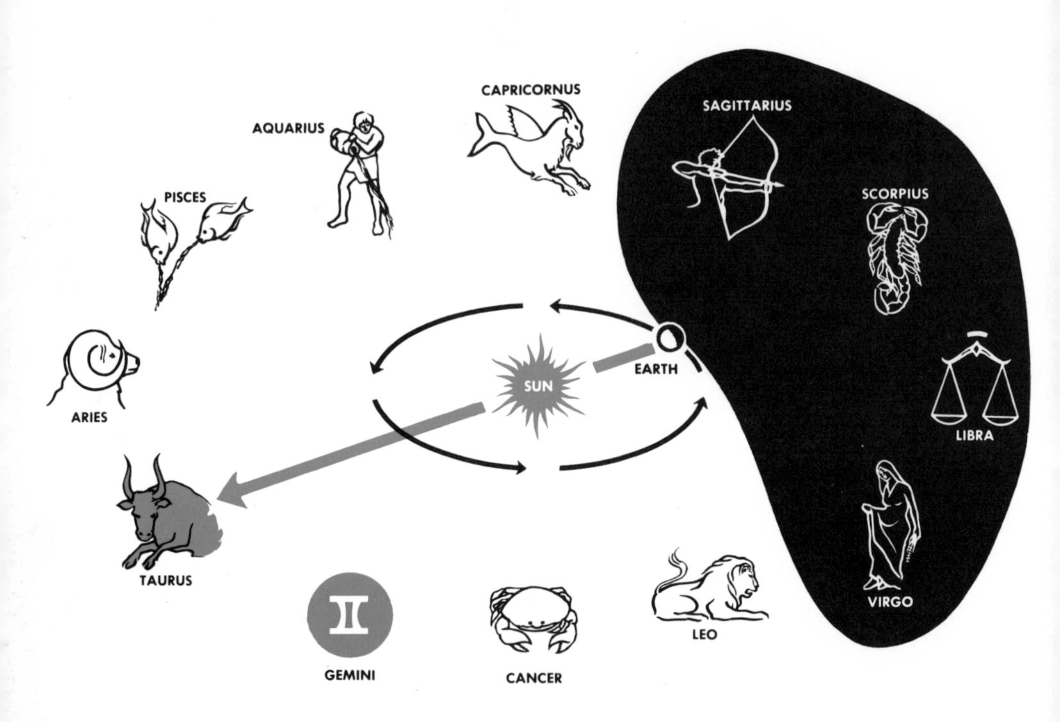


The sun is in the direction of Taurus
from May 13 to June 18.
Gemini is the astrological sign
from May 21 to June 21.

Virgo is in the evening sky
along with Centaurus
and the Southern Cross.

Hunting Dogs, Berenice's Hair, Cup, Crow, and Fly
are near the meridian
early in the evenings at mid-month.

Omega Centauri is interesting
to explore with binoculars,
and the Realm of Galaxies with a telescope.

**VIRGO, THE VIRGIN**, Goddess of Justice and Purity, is a big constellation with one bright star. The sun spends 43 days in this zodiacal group, from mid-September to the end of October, and is in Virgo for the autumnal equinox.

Spica is a hot blue-white star six hundred times as bright as the sun, at a distance of 220 light-years. It is a grain of wheat, in some celestial representations, that Virgo holds in her hand.

Spica

Beta Centauri

Alpha Centauri

**CENTAURUS, THE CENTAUR**, one of the biggest constellations, is best known for its two bright stars that are pointers to the Southern Cross.

The brighter of the pointers, Alpha Centauri, is the third brightest star in the sky. Only Sirius and Canopus are brighter. Alpha Centauri, also known as Rigil Kentaurus, is the closest star to the sun—4.3 light-years distant. It is a double that can be separated into its components with a small telescope.

Half-man, half-horse, centaurs were the only monsters of old capable of doing good deeds. Some were rude creatures who got drunk at weddings and broke the furniture. But the wisest and most just of the lawless race, Chiron (Centaurus), was a kindly centaur who taught the arts of war and peace to the heroes of Greek mythology.

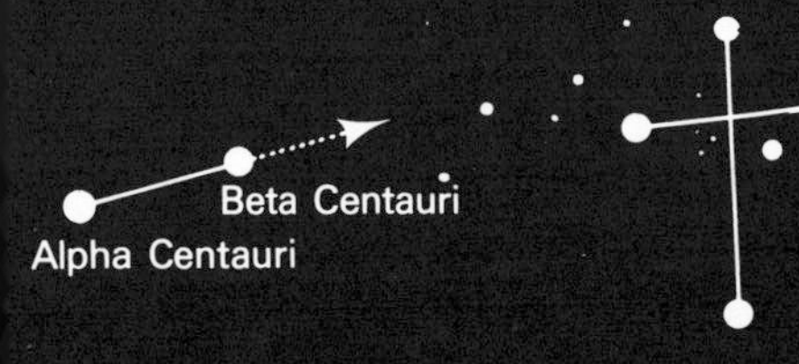


**CRUX, THE SOUTHERN CROSS** is a small, bright group of stars set in the Milky Way.

The Southern Cross is visible from the extreme southern parts of Florida, Texas, and any of the islands of Hawaii. The Cross is low on the horizon as seen from the northern tropics—the region on earth from which you can truly see from North Star to Southern Cross.

Crux circles the southern sky, its staff pointing almost directly toward the south celestial pole. However, there is no "south star" as there is a "north star" (Polaris).

May is the time for the best view of the sky from North Star to Southern Cross. Leo moves to the west with the Realm of Galaxies trailing him. And Virgo, the longest zodiacal constellation, spanning 52 degrees of sky, moves to the center.

Virgo is associated with the harvest in many cultures. She holds the Grain of Wheat or the Ear of Corn in her hand. "In Virgo are contayned infirmities, thinges contrary to health, maydes, lying, unrighteousness," says the Arcandam book of 1592.

CRATER, THE CUP is a two-handled pot in England. Elsewhere, it is called the Cup of Apollo, the Cup of Bacchus, the Cup of Achilles, the Cup of Hercules—and even the Cup of Medea, the sorceress who helped Jason get the Golden Fleece.

CANES VENATICI, THE HUNTING DOGS chase the Great Bear around the north pole—greyhounds on the leash of Boötes. The brightest star in the group, *Cor Caroli,* Charles' Heart, was named in honor of Charles I of England by Halley, the astronomer who correctly identified the period of the comet that now carries his name.

COMA BERENICES, BERENICE'S HAIR is a shimmering patch of 5th magnitude stars—an open cluster at a distance of 270 light-years. Berenice, a queen of Egypt with amber tresses, dedicated her hair to the Goddess of Beauty in order to ensure her husband's safe return from a dangerous voyage.

The Realm of Galaxies is in the direction of the north galactic pole, where the view of deep space is best. Light from the galaxies there takes millions of years to reach us. It is a thrilling experience to look at this region of deep space with your eye to the telescope, and to photograph the universe as it was 150 million years ago, at the time dinosaurs were roaming the earth.

Apollo used CORVUS, THE CROW as a spy to check on his beloved Coronis. Red Bird, Storm Bird, Raven, and Desert Bird are other names for this lopsided square of stars. Far to the south is MUSCA, THE FLY, buzzing about the feet of Centaurus.

Less than fifty years ago the "Coal Sack" in the Southern Cross was thought to be "an opening into the awful solitude of unoccupied space . . . a curious vacancy through which we seem to gaze out into an uninterrupted infinity." Today the radio telescope shows the Coal Sack to be a dark cloud of hydrogen gas 450 light-years away, lying in front of a shining background of stars. Corsali wrote of the Southern Cross in 1517, "This Cross is so fayre and beautiful that none other heavenly sygne may be compared to it."

CEPHEUS
CAMELOPARDALIS
PERSEUS
KIDS
Capella
AURIGA
M 38
M 37
LYRA
Vega
Eltanin
DRACO
LITTLE DIPPER
URSA MINOR
Polaris
Kochab
M 81
Thuban
Dubhe
Megrez
BIG DIPPER
Merak
Alcor
Alioth
Mizar
Phecda
URSA MAJOR
LYNX
M 13
HERCULES
Alkaid
M 51
Castor
GEMINI
Pollux
CORONA BOREALIS
BOÖTES
Cor Caroli
CANES VENATICI
LEO MINOR
Ecliptic
Gemma
M 3
Algenubi
M 44
BEEHIVE
CANCER
SERPENS CAPUT
SICKLE
Arcturus
COMA BERENICES
LEO
CANIS MINOR
Procyon
Denebola
Regulus
Realm of Galaxies
VIRGO
September Equinox
SEXTANS
MONOCEROS
Zubeneschamali
Alphard
LIBRA
Zubenelgenubi
Spica
CRATER
Gienah
CORVUS
HYDRA
ANTLIA
PYXIS
Menkent
CENTAURUS
Suhail
LUPUS
OMEGA CENTAURI
VELA
SOUTHERN CROSS
Gacrux
COAL SACK
Hadar
Rigil Kentaurus
Acrux
CIRCINUS
MUSCA
CARINA
May

# JUNE

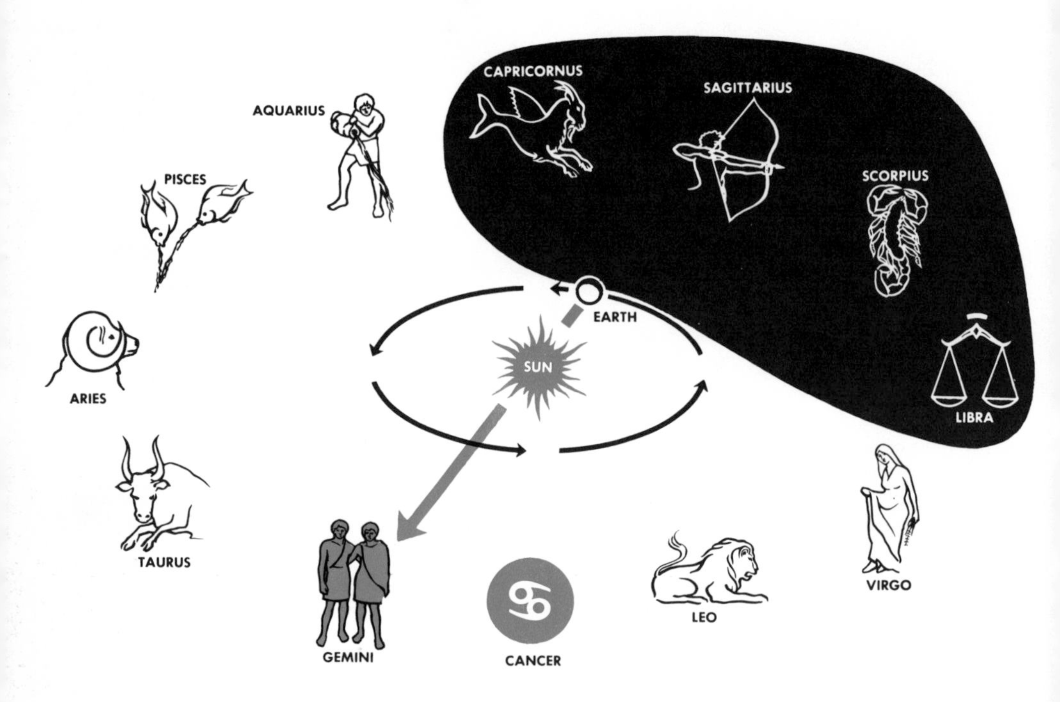


The sun is in the direction of Gemini
from June 18 to July 18.
Cancer is the astrological sign
from June 22 to July 22.

Libra is in the evening sky
along with Boötes and Ursa Minor.

Wolf and Compasses are near the meridian
early in the evenings at mid-month.

M 3, M 51, Rigil Kentaurus, Arcturus,
and the Omega Centauri Cluster
are objects of interest.

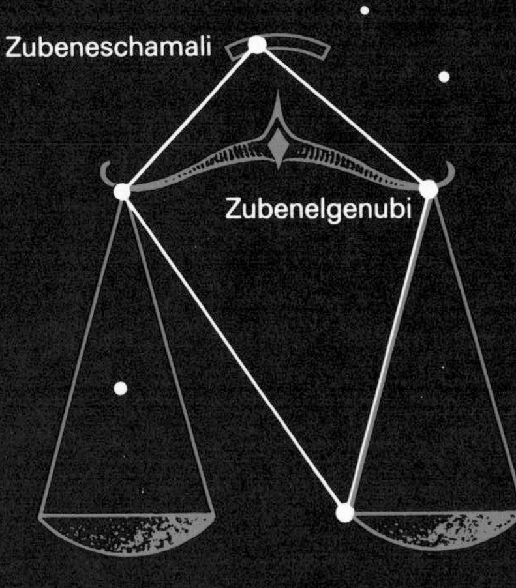


**LIBRA, THE SCALES** is the only constellation in the Circle of Animals that is not a life form. Long ago its stars were the pincers of Scorpius, but later the Romans separated them from Scorpius to form the last of the twelve constellations along the path of the sun. Now Libra represents the Scales that the Goddess of Justice, Virgo, holds in her hands.

The brightest stars in Libra have the wonderful cabalistic-sounding names of Zubenelgenubi and Zubeneschamali.

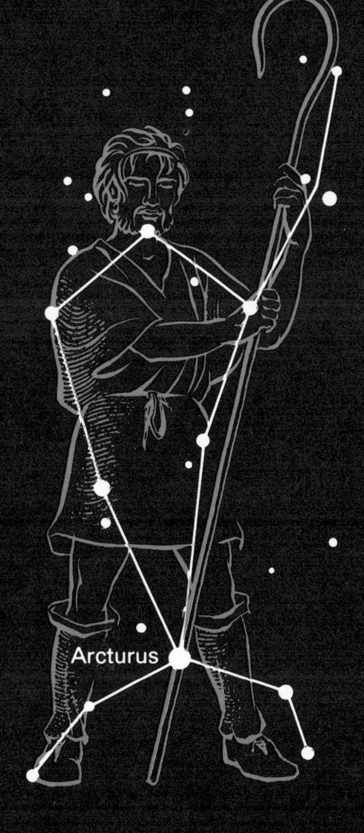


**BOÖTES, THE HERDSMAN** (pronounce both o's) is a kite-shaped group. Ahead of the Herdsman (also called the Bear Driver) are the hunting dogs, Canes Venatici, on the trail of the Great Bear.

Orange-red Arcturus, the brightest star in the northern hemisphere, was a beacon in the sky for the Polynesians, marking the end of the journey in their voyages from Tahiti to Hawaii. Its zenith passage takes it over Vietnam, Rangoon, Bombay, Mecca, Havana, Mexico City, and Honolulu.

Arcturus is a giant, 22 times the sun's diameter, 83 times as bright but slightly cooler. Its density is so low—1/5,000 that of water—that it's practically a vacuum; but even that is very dense compared to the emptiness of space around it.

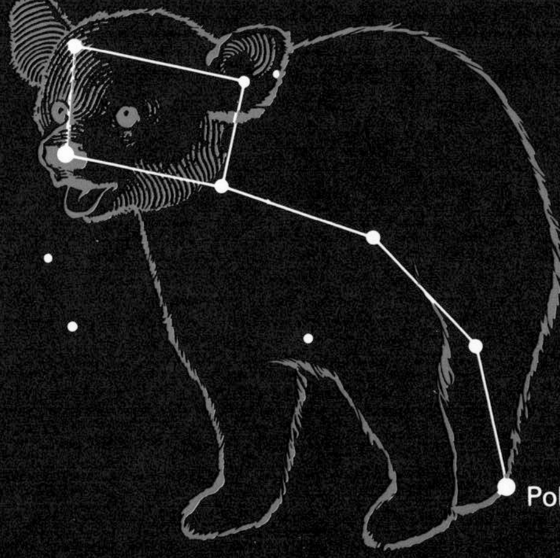


**URSA MINOR, THE LITTLE BEAR**, commonly known as the Little Dipper, looks more like a saucepan than a dipper, for its handle curves the wrong way.

Polaris is now within a degree of being directly over the earth's north pole. Five hundred light-years distant, the pole star is a double that can be separated into its components with a 3-inch telescope.

Draco, the Dragon now guards the Little Dipper, just as in ancient times he guarded the Tree of Golden Apples. Boötes and his dogs chase the Bear toward the setting sun, and Libra, the symbol of equality, moves to the center of the sky.

"Whoſo iſ borne in yat syne sal be an ille doar and a traytor," said a 14th-century manuscript of Libra. Writers of more sanguine temperament called it the House of Venus, for it was in Libra that the Goddess appeared at Creation. Autumnal equinox once was in Libra. Beam and balance represented in Libra's symbol ♎ express the equality of day and night at equinoctial times—"then Day and Night are weigh'd in Libra's Scales." Precession has now made that meaning less precise, for the autumnal equinox has moved to Virgo.

The "fixed stars" of ancient times do move, and, given time enough, the heavens would eventually be unrecognizable from our star maps. Arcturus is relatively close to the earth—36 light-years—and moves rapidly compared to other stars, so rapidly that in 8,000 years it will have moved five degrees, or about the angular distance between the Pointers of the Big Dipper. Sixty thousand years from now it will be close to Spica, which will have moved only one degree, being a much more distant star.

M 3, a globular cluster 45,000 light-years away, is between Arcturus and Cor Caroli. The Whirlpool Nebula M 51 appears as two glowing clouds in a 3-inch telescope; larger instruments reveal a tight spiral connected to a smaller group.

LUPUS, THE WOLF, is an old constellation. As *Victima Centauri,* he is the offering that Centaurus makes to the gods upon the altar, Ara. CIRCINUS, THE COMPASSES is a faint group noted for its doubles ranging in color from yellow to bright red.

Alpha Centauri, or Rigil Kentaurus, the "Centaur's Foot," was worshipped by the people along the Nile. Nine Egyptian temples built in the 3800-2500 B.C. period were oriented to the first visible emergence from the sun's rays of this "object of splendor." Alpha Centauri is the closest star to the sun, at 4.3 light-years—250,000 times the earth-sun distance. This third brightest star in the sky is a double, the two stars revolving around each other in 80 years. The brighter of the pair is almost the same luminosity, temperature, size, and mass as the sun.

Delta Cephei
VARIABLE
CAMELOPARDALIS
CEPHEUS
M39
Alderamin
Polaris
Deneb
M 81
LYNX
Castor
URSA MINOR
GEMINI
LITTLE DIPPER
Kochab
Pollux
DRACO
BIG DIPPER
Dubhe
Eltanin
Megrez
Merak
Thuban
URSA MAJOR
Phecda
LYRA
Vega
Alioth
M 57
Beta Lyra
Alcor
Mizar
M 51
Alkaid
LEO MINOR
Algenubi
M 13
HERCULES
Cor Caroli
SICKLE
CANCER
CORONA BOREALIS
CANES VENATICI
BOÖTES
M 3
Gemma
LEO
COMA BERENICES
Regulus
Rasalhague
Ras-Algethi
Arcturus
Denebola
Ecliptic
SERPENS CAPUT
Realm of Galaxies
SEXTANS
OPHIUCHUS
VIRGO
September Equinox
CRATER
Zubeneschamali
Spica
LIBRA
Sabik
Zubenelgenubi
CORVUS
Gienah
Ecliptic
Dschubba
Antares
HYDRA
SCORPIUS
Menkent
M6
M7
ANTLIA
LUPUS
CENTAURUS
OMEGA CENTAURI
NORMA
SOUTHERN CROSS
VELA
Gacrux
ARA
Hadar
COAL SACK
Rigil Kentaurus
Acrux
CARINA
TRIANGULUM AUSTRALE
CIRCINUS
MUSCA
Atria
June

# stars of SUMMER

**Summer positions of stars near the North Star.**

Two giants, two birds, two centaurs, two crowns, a musical instrument, a scorpion, and a dragon are in the summer sky.

Arcturus, a star of early summer, was a beacon in the sky for the ancient Polynesians, for its zenith passage marked the latitude of the islands they were sailing for in their long voyages from Tahiti and the Marquesas to Hawaii. Each night Arcturus reaches the meridian four minutes earlier. By late summer it twinkles in the western twilight, soon to be gobbled up by the sun.

Tethered to Polaris, the Little Dipper swings as far south as it can go. Draco cradles its bowl. Leo sets, and the Summer Triangle moves to the center of the sky.

The Summer Triangle is a large, right-angle triangle spanning the Milky Way. Often called the Navigator's Triangle, its three bright stars are among the first to be seen in the evening—important for the navigator in getting a "fix" on them and measuring their altitudes while there is still enough twilight to see the horizon. Deneb, Vega, and Altair (the points of the Triangle) are three stars of three different constellations—the Swan, the Harp, and the Eagle.

Vega might well be called the "Central U.S. Star", for its zenith passage takes it over Washington, St. Louis, Denver and San Francisco. But it might also be called the "Capital Star": it passes over Tokyo, Peking, Tehran, Ankara, Athens, Rome, Madrid, Lisbon, and Washington.

The Milky Way is a faint band of light arching northeast-to-southwest across the heavens. It is at its best during the summer, for at this season we're looking toward its brightest part—the galactic center. So plentiful, so distant, and so close to the same direction are the stars in this region that we see them as faint clumps, clouds, and clusters.

For observers in the more southern latitudes, Scorpius dominates the summer sky just as Orion does the winter. The two are associated in legend even though they're in opposite portions of the sky. Orion boasted of his strength, saying that neither man nor beast could overcome him. Jupiter heard his boasting. Instead of hurling a lightning bolt at Orion—at this mortal who considered himself equal to the gods—Jupiter sent Scorpius to bite him on the heel. Mortal Orion, succumbing to the sting of death, is now immortalized in the sky along with his victor. But they're incompatible—180 degrees apart.

# JULY

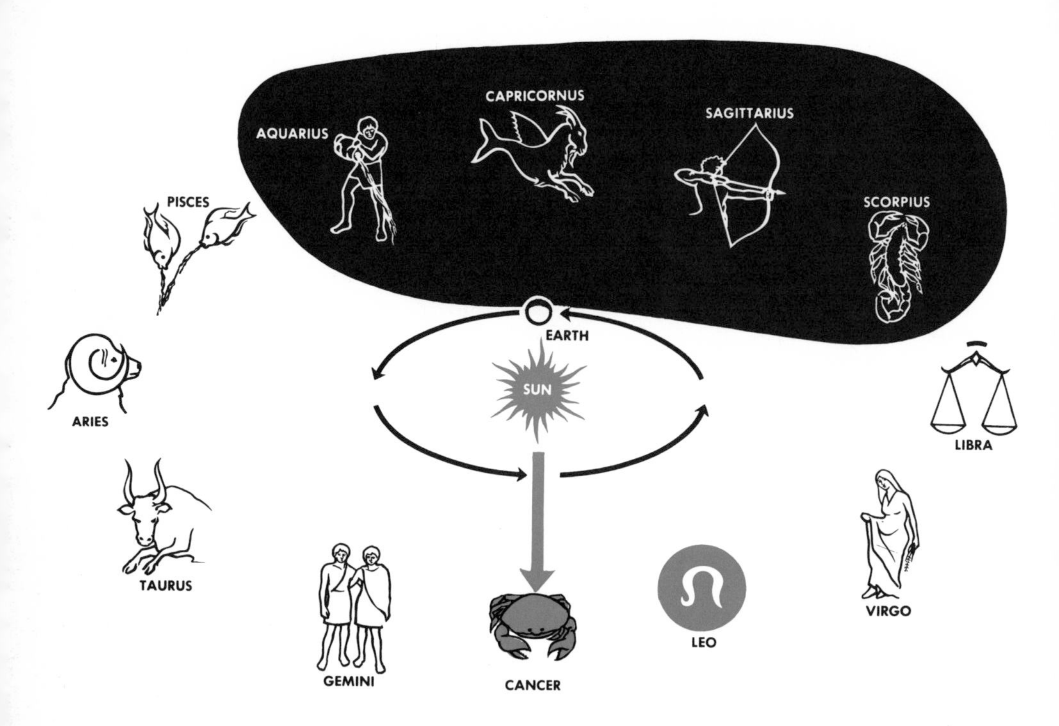


The sun is in the direction of Cancer
from July 18 to August 7.
Leo is the astrological sign
from July 23 to August 22.

Scorpius is in the evening sky
along with
Hercules, Draco, Ophiuchus, and Serpens.

Level and Southern Triangle
are near the meridian
early in the evenings at mid-month.

The Great Cluster M 13 in Hercules
is a fascinating object in binoculars.

**SCORPIUS, THE SCORPION**, immersed in the Milky Way, stands out clearly from the distant background of faint star clouds.

Giant red Antares, four-hundred times the diameter of the sun, burns fiercely as the heart of Scorpius. This "Rival of Mars" lies close to the ecliptic, and when Mars occasionally draws close to it, the two do rival each other in color and brilliance.

Three stars mark Scorpius' head, red Antares his heart, and a sharply-curving group (many of which are doubles), his tail and stinger.

Antares

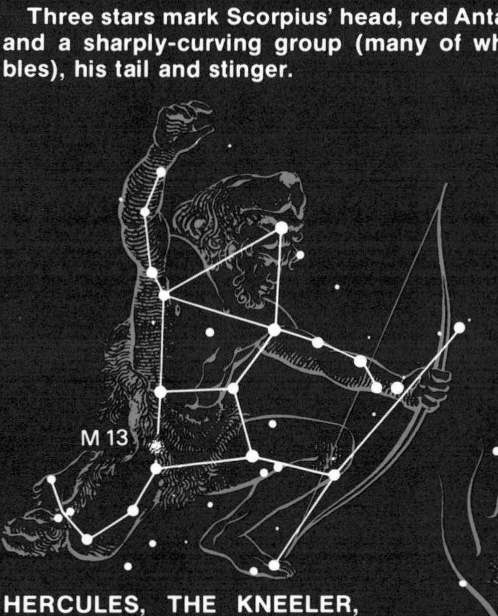


**HERCULES, THE KNEELER**, was famous in mythology for his twelve labors that demanded valor and heroism.

The hazy patch on his western side is the Great Cluster, M13—a group of 100,000 stars at a distance of 36,000 light-years.

Eltanin

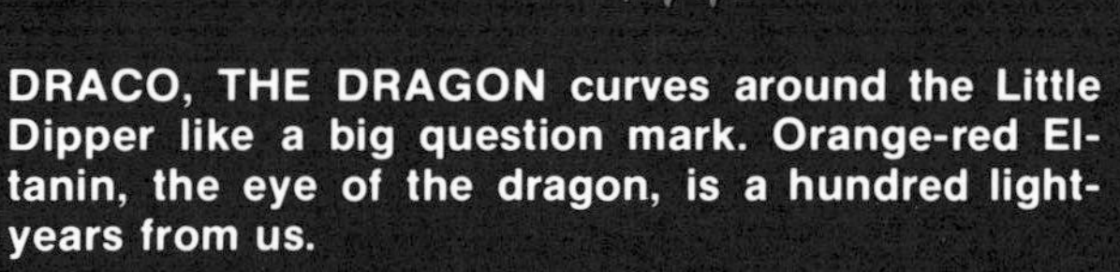

**DRACO, THE DRAGON** curves around the Little Dipper like a big question mark. Orange-red Eltanin, the eye of the dragon, is a hundred light-years from us.

**OPHIUCHUS, THE SERPENT HOLDER**, son of Apollo and pupil of the centaur Chiron, became a successful physician. The snake's periodic shedding of its skin is an emblem of health signifying the renewal of life.

The sky is busy this month with Hercules beating on the head of the Dragon, and with Ophiuchus standing on Scorpius as he wrestles the Serpent. Rasalgethi and Rasalhague, the heads of the two giants, are close to each other. And the celestial scene seems an allegory of the triumph of good over evil.

Ophiuchus was a doctor whose professional competency was a deterrent to his own well-being. So accomplished was he in the art of making people well that Pluto complained to Jupiter that there was not enough business in Hades. Jupiter solved the problem by elevating the good doctor into the heavens.

Hercules had twelve great labors to perform. The first was getting the skin of the Nemean Lion, whom Leo represents. Slaying Hydra was another. He successfully cleaned the Augean Stables, which had remained uncleaned for 30 years, by diverting two great rivers through them. He took the jeweled belt of the Queen of the Amazons, and he successfully brought back the three-bodied oxen of Geryon as well. As he traveled to the ends of the earth in search of the Golden Apples of the Hesperides he gave Atlas a little rest by holding up the heavens for a while.

Draco, the Dragon guarded the Golden Apples. He lived in a beautiful garden far beyond the mountains on which powerful Atlas stood. Draco coiled about the Tree of Golden Apples, never sleeping. A band of nymphs danced around him, singing entrancing songs to keep him awake. His eyes sparkled. Atlas located the Golden Apples and dealt with Draco while Hercules held up the heavens. Relieved of his burden and possessing the prize, Atlas decided to take the Apples back to Eurystheus himself. But Hercules tricked him into taking the heavens back again while he put a pad beneath his shoulder. Slow-witted Atlas resumed his burden and Hercules returned with the Golden Apples.

The stars in Hercules' body form a figure that resembles butterfly wings. Between the two western stars in the butterfly is the Great Cluster M 13, visible to the unaided eye under good viewing conditions. A telescope reveals dark lanes and star streams in this close globular cluster, 22,000 light-years from earth.

NORMA, THE LEVEL, in the dark part of the Milky Way, was originally part of *Norma et Regula* (Level and Square), but the Square has become TRIANGULUM AUSTRALE, THE SOUTHERN TRIANGLE. Second-magnitude Atria, 21 degrees from the south celestial pole, is the Triangle's brightest star, 82 light-years from us.

CASSIOPEIA
Caph
MEDA
CAMELOPARDALIS
LYNX
Delta Cephei
VARIABLE
CEPHEUS
Polaris
M 81
URSA MAJOR
LACERTA
Alderamin
URSA MINOR
Dubhe
Merak
M 39
LITTLE DIPPER
Kochab
BIG DIPPER
Phecda
Megrez
Deneb
Thuban
Alioth
LEO MINOR
CYGNUS
DRACO
Alcor
Mizar
M 51
Eltanin
Alkaid
CANES VENATICI
LYRA
Vega
Cor Caroli
Beta Lyra
M 57
M 13
M 3
LEO
Albireo
M 27
CORONA BOREALIS
BOÖTES
COMA BERENICES
Denebola
HERCULES
SAGITTA
Gemma
INUS
Arcturus
Realm of Galaxies
Altair
Ras-Algethi
Rasalhague
SERPENS CAPUT
September Equinox
VIRGO
OPHIUCHUS
SERPENS CAUDA
M 11
Zubeneschamali
Ecliptic
Spica
SCUTUM
LIBRA
Zubenelgenubi
CORVUS
Sabik
M 17
M 23
Ecliptic
M 20
Dschubba
M 8
December Solstice
HYDRA
Nunki
Antares
TEAPOT
SCORPIUS
M 6
M 7
Shaula
Menkent
GITTARIUS
Kaus Australis
LUPUS
CORONA AUSTRALIS
CENTAURUS
OMEGA CENTAURI
NORMA
TELESCOPIUM
ARA
SOUTHERN CROSS
Gacrux
Hadar
Rigil Kentaurus
COAL SACK
Acrux
PAVO
CIRCINUS
TRIANGULUM AUSTRALE
Atria
July

# AUGUST

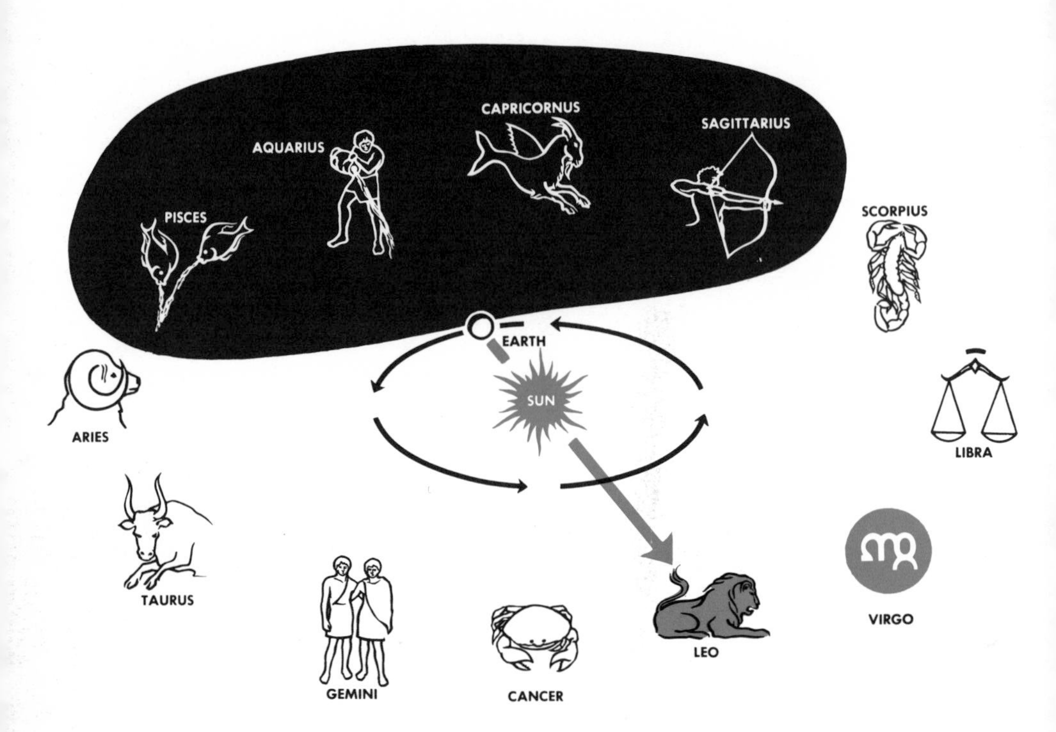


The sun is in the direction of Leo from August 7 to September 14. Virgo is the astrological sign from August 23 to September 22.

Sagittarius is in the evening sky along with Lyra, Corona Borealis, and Corona Australis.

Shield and Altar are near the meridian early in the evenings at mid-month.

M 6, M 7, M 23, M 20, M 17, and M 11 are interesting clusters and nebulae. The Ring Nebula in Lyra, M 57, is a good object in a telescope.

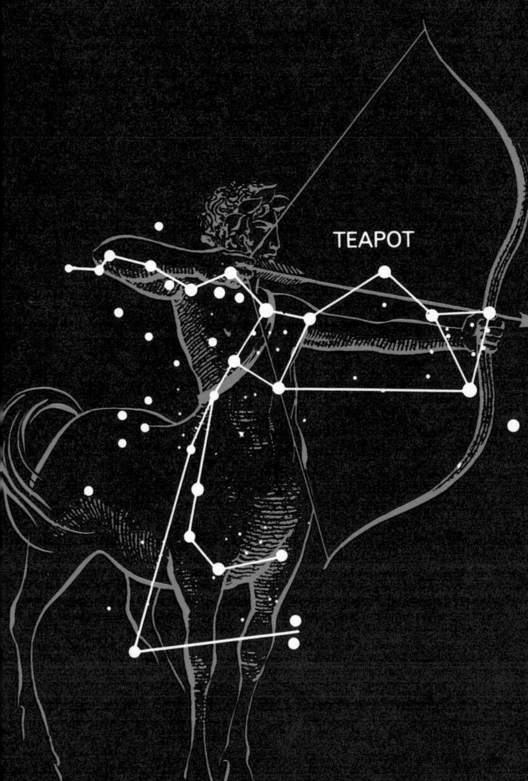


**SAGITTARIUS, THE ARCHER** (and a centaur as well) aims an arrow at Scorpius.

The stars in Sagittarius are not bright (2nd and 3rd magnitude) but their spacing gives the impression of fine strength, like a webbed network of steel. Once you find it, you'll look for it again and again in the summer sky.

The upper part of Sagittarius also looks like a celestial teapot, spilling its contents on the tail of Scorpius as it moves beyond the meridian.

The center of our Galaxy lies in the direction of Sagittarius, 30,000 light-years distant. Strong gravitational and radio waves emanating from that center indicate the magnitude of the turmoil within.

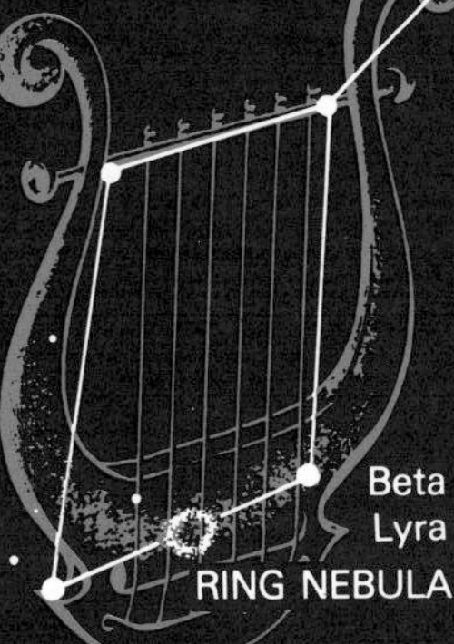


**LYRA, THE LYRE** represents the stringed instrument that Apollo gave to Orpheus. The harp gave forth such beautiful music in the hands of Orpheus that it had the power to tame wild beasts and to still rivers.

Blue Vega, 27 light-years distant, is about the same brightness as Arcturus and Capella. The only brighter stars—Sirius, Canopus, and Rigil Kentaurus—are all in the southern hemisphere.

Beta Lyra is an eclipsing variable whose variations in light can be observed with the unaided eye.

The famous Ring Nebula is the remains of a star that, having puffed off its hydrogen, is collapsing into a white dwarf.

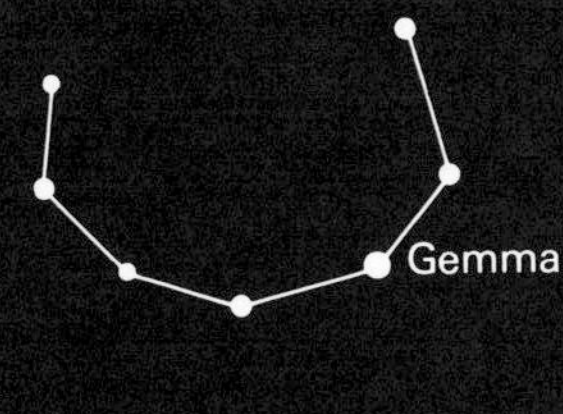


**CORONA BOREALIS, THE NORTHERN CROWN** is a semicircle of faint stars. Even though the individual stars are not bright, the constellation is easy to see because of its closely spaced stars. Sparkling Gemma is the brightest in the Crown.

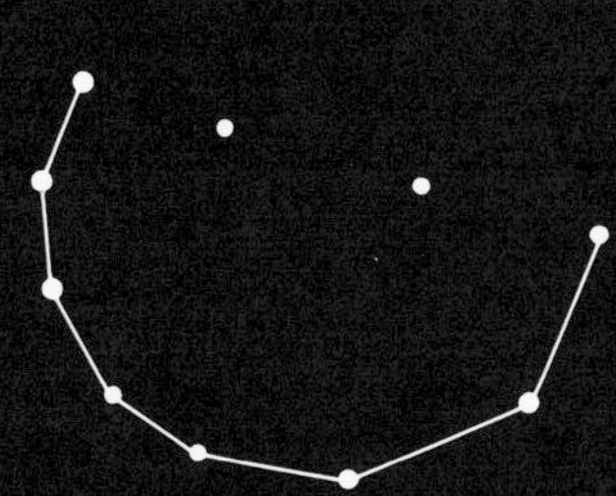

**CORONA AUSTRALIS, THE SOUTHERN CROWN** is also a very faint group; it lies east of the tail of the Scorpion and below the Archer, and is visible from the southern extremities of the United States.

A dragon's head, an arrowhead, a serpent's tail and a scorpion's stinger are on the meridian. At the place of the December solstice is a group of diffuse nebulae and clusters. The central and brightest part of the Milky Way is there, too, in the direction of Sagittarius.

"In Sagittarius are expressed long journies, religion, wisedome, philosophie, wrytinges, bookes, news, and interpretation of dreams, great wonders, much honour and joy," wrote Dr. Arcandam in 1592. He also went on to state that a man born under this sign would be thrice wedded, and very fond of vegetables.

SCUTUM, THE SHIELD is just above the ecliptic, right in the Milky Way. Shaped like an elongated diamond, Scutum is small and faint, and hard to see because of the rich star field behind it. Three hundred years ago, it was named *Scutum Sobieski,* honoring a king of Poland who saved his country from invasion.

Near the upper point of the Shield is the M 11 cluster, rich and concentrated, just visible to the unaided eye. Herschel in 1681 called it "a glorious object." It's a triangular patch of hundreds of stars 5,500 light-years distant.

The Omega Nebula, M 17, is also known as the Horseshoe Nebula and the Swan Nebula. It lies at a distance of 3,000 light-years from earth. Young, hot stars within the nebula produce strong ultraviolet radiation that excites the cloud into brilliance.

The Trifid Nebula (M 20) and the Lagoon Nebula (M 8) are also clouds of hydrogen lighted in the same way. The Trifid Nebula, 3,500 light-years distant, has three dark lines , or "rifts" that meet at the center. The brighter Lagoon Nebula, a thousand light-years farther into space, is large in diameter and can be seen with the unaided eye. Nearby is M 23, an open cluster (like Coma Berenices) of 120 stars in a curving line.

Just above the tail of Scorpius are M 6 and M 7—brilliant and extensive open clusters but low in the sky. M 6 lies at a distance of 1,500 light-years; the closer M 7 is only half as far.

Beneath "the glowing sting of Scorpius" is ARA, THE ALTAR—the Altar of the Centaur, the Altar of Dionysus, the Altar on which perfumes were burned for the dead . . . an Incense Pan, the Brazier, the Censer, the Hearth. Smoke and flames rising from the Altar drift northward to form the Milky Way. Ara, so far to the south, spends "but little time aloft and it quickly speeds beneath the western sea," according to Aratus' observations.

The Ring Nebula in Lyra, M 57, is visible in a 12-inch telescope. It's a spherical shell of gas surrounding a central hot blue star, 5,000 light-years from us.

Almach
M 103
Ruchbah
Navi
Schedar
CASSIOPEIA
M 31
Caph
CAMELOPARDALIS
M 81
Dubhe
Merak
URSA MAJOR
LEO MINOR
Polaris
BIG DIPPER
Phecda
Megrez
URSA MINOR
LITTLE DIPPER
Kochab
Alioth
CANES VENATICI
CEPHEUS
Delta Cephei VARIABLE
Alderamin
Thuban
Mizar
Alcor
M 51
Cor Caroli
DRACO
Alkaid
LACERTA
M 39
Deneb
Eltanin
M 3
COMA BERENICES
CYGNUS
BOÖTES
M 13
LYRA
Vega
CORONA BOREALIS
Beta Lyra
M 57
HERCULES
Gemma
Arcturus
VULPECULA
M 27
Albireo
M 15
DELPHINUS
SAGITTA
Enif
EQUULEUS
Altair
Ras-Algethi
Rasalhague
SERPENS CAPUT
VIRGO
AQUILA
OPHIUCHUS
Zubeneschamali
SERPENS CAUDA
Ecliptic
Zubenelgenubi
M 11
LIBRA
SCUTUM
M 17
Sabik
M 23
Dschubba
M 20
M 8
December Solstice
Nunki
Antares
CAPRICORNUS
SAGITTARIUS
TEAPOT
SCORPIUS
M 7
M 6
Shaula
Kaus Australis
CORONA AUSTRALIS
LUPUS
MICROSCOPIUM
TELESCOPIUM
NORMA
ARA
INDUS
PAVO
CIRCINUS
TRIANGULUM AUSTRALE
August

# SEPTEMBER

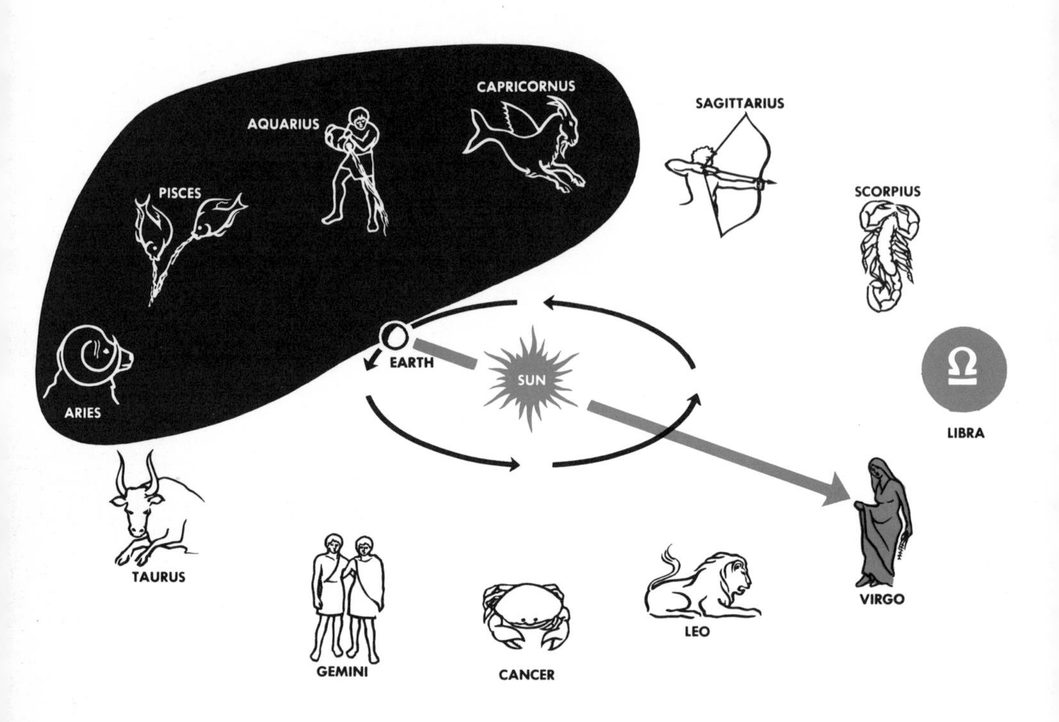


The sun is in the direction of Virgo
from September 14 to October 29.
Libra is the astrological sign
from September 23 to October 23.

Capricornus is in the evening sky
along with Cygnus and Aquila.

Fox, Arrow, Dolphin, Microscope, and Telescope
are near the meridian
early in the evenings at mid-month.

The Ring Nebula in Lyra
and the yellow-blue double star Albireo
are interesting objects.

**CAPRICORNUS, THE GOAT** has the shape of a wide arrowhead—a triangle with a sagging hypotenuse. The fish-tailed, sea-going Goat is the first of the three "watery" constellations of the zodiac—all of which are hard to see.

The sun in ancient times was in the direction of Capricornus at its extreme southern position; the sign of the Goat was the "Southern Gate of the Sun."

Deneb

Albireo

**CYGNUS, THE SWAN** flies down the Milky Way, head and neck outstretched, wings outspread.

Cygnus represents Orpheus the Musician, in one legend, placed in the sky to be near his magic harp, Lyra. But in another legend, the Swan is the form that Jupiter assumed so that he could visit his beloved Leda without being recognized by his jealous wife Juno.

Deneb, the white super-giant in the Swan's tail, is a tremendous star: brighter than 60,000 suns and a billion times farther. The sun would not be visible to us at such a distance.

Albireo at the head of Cygnus is a double star. One of its components, seen through the telescope, is sapphire-blue in color; the other, topaz-yellow.

**AQUILA, THE EAGLE** flies along the Milky Way toward Cygnus. Jupiter changed himself into the Eagle to take Ganymede, a Trojan lad, to be his cupbearer.

Bright Altair is easy to recognize for its two guide stars, one on each side. Light from Altair takes 17 years to reach the earth.

Altair, Deneb, and Vega make a bright right triangle that spans the Milky Way.

Altair

**SAGITTA, THE ARROW** lies between the two birds, symbolizing one of Hercules' labors—the encounter with the Stymphalian Birds.

The giants Hercules and Ophiuchus are three hours past the meridian and going toward their setting places, and the Winged Horse and 'watery' constellations are rising high. The Little Dipper streams westward from Polaris, and the Navigator's Triangle straddles the meridian.

Capricornus is the fish-tailed animal into which Pan was changed in order to escape the terrible monster Typhon. Goat-faced Pan, terrified by Typhon's approach, fled—in panic—becoming fishified in his plunge into the Nile. Capricornus was the Gate of the Gods through which the souls of mortals ascended into the heavens. And of the powers of the sign, the Almanac of 1386 says, "Whoſo iſ borne in Capcorn shal be ryche and wel lufyd."

Hevelius crowded VULPECULA, THE FOX between Swan and Arrow. He admired the fox: "Such an animal is very cunning, voracious, and fierce." The Dumbbell Nebula, M 27, is its most notable object—a planetary nebula 980 light-years distant, but spread over such a large area that it is a difficult object in a small telescope.

Flying toward the east between Fox and Eagle is SAGITTA, THE ARROW. It's the Eagle's talons in some representations. But in most, an arrow—the Arrow that Hercules shot toward the Stymphalian Birds, or the Arrow that Apollo used on the Cyclops, or even Cupid's Arrow. But if it is the Arrow of Sagittarius, as some say, then it has not only strayed badly but even reversed direction.

DELPHINUS, THE DOLPHIN is a diamond in the sky with a tail—an emblem of philanthropy because of the Dolphin's devotion to its young. As *Persuasor Amphitrites,* the affable Dolphin persuaded the Goddess of the Sea (Amphitrites) to become the wife of Neptune.

MICROSCOPIUM and TELESCOPIUM are inconspicuous groups of faint stars south of Capricornus and Sagittarius. La Caille formed them, along with the Chemist's Furnace, Sculptor's Chisels, Air Pump, Carpenter's Square, and Painter's Easel. Thirty years later, Bode put the Electric Machine, *Machina Electrica,* in the sky south of Cetus, but it did not last. La Lande formed *Felis,* the Cat (seen on old charts as Katze and Gatto) near Antlia, saying, "I am very fond of cats. I will let this figure scratch on the chart. The starry sky has worried me quite enough in my life, so that now I can have my little joke with it." But the joke itself has since been scratched from celestial charts, and Felis no longer appears in the sky.

CAMELOPARDALIS
Megrez
URSA MAJOR
Alioth
CANES VENATICI
Mirfak
PERSEUS
Polaris
Mizar
M51
Cor Caroli
Kochab
Thuban
Alcor
DOUBLE CLUSTER
M34
COMA BERENICES
M 103
CASSIOPEIA
Ruchbah
URSA MINOR
LITTLE DIPPER
Alkaid
Navi
Almach
M3
Schedar
DRACO
CEPHEUS
Caph
Delta Cephei
VARIABLE
Alderamin
BOÖTES
M31
Mirach
ANDROMEDA
LACERTA
Eltanin
Arcturus
M13
CORONA BOREALIS
Gemma
M 39
Deneb
Alpheratz
Vega
HERCULES
LYRA
M 57
Beta Lyra
Scheat
GREAT SQUARE
CYGNUS
Albireo
SERPENS CAPUT
VULPECULA
PEGASUS
M 27
Ras-Algethi
Markab
DELPHINUS
SAGITTA
Rasalhague
M 15
Altair
CIRCLET
Enif
EQUULEUS
WATER JAR
AQUILA
OPHIUCHUS
SERPENS CAUDA
M 11
AQUARIUS
SCUTUM
M 17
M 23
Sabik
M 20
Ecliptic
CAPRICORNUS
Nunki
M 8
December Solstice
Antares
SAGITTARIUS
TEAPOT
PISCIS AUSTRINUS
M6
M7
Shaula
Kaus Australis
SCORPIUS
CORONA AUSTRALIS
MICROSCOPIUM
TELESCOPIUM
GRUS
Al Nair
ARA
NORMA
INDUS
PAVO
TUCANA
September

# stars of AUTUMN

Autumn positions of stars near the North Star.

The sun crosses the sky lower and lower each autumn day; the nights grow longer.

Late in December the sun enters the constellation Sagittarius, and is just above the Archer's Bow on the shortest day of the year. Here, at the southern limit of the sun's sweep, is the winter solstice where the sun seems to stand still for a while, resting for its six-months' journey northward.

Autumn, like the spring season, is an in-between time for stargazers. The brilliant stars of summer are far to the west, and the brilliant stars of winter are beginning to rise. Dull and drab in comparison, the autumn sky has a beauty of its own, reflected in lore and legend.

Autumn, though an off-season for bright stars, is a time of wonderful creatures, strange stars, and celestial characters. For this is the time of the Royal Family—the Vain Queen, Sorrowful King, Enchained Lady, and Modest Hero; the time of the Snake-Haired Monster, Winged Horse, and ravenous Sea Beast; the time of the "watery" constellations—Water-Carrier, River, Sea-Going Goat, Dolphin, Wading Bird, and three famous Fishes—faint star groups scattered over a huge background like tiny islands in a dark sea.

The Big Bear has gone to his den beyond the North Star; and Cassiopeia is at her best. Antares, Kaus Australis, and Fomalhaut sweep across the southern sky, and Aquarius shoulders his Water Jar.

Autumn is also the time of the pulsating star, Delta Cephei; the time of the sinister winking Demon Star, Algol; and the time of the disappearing star, Mira.

Autumn is the time when we look the deepest into space with the unaided eye and have an object to see—a galaxy twice as big as ours and two million light-years distant. And it gives us pause to think that the light reaching us tonight from that great spiral had completed 99.7% of its journey at the time the Egyptians were building their Pyramids.

# OCTOBER

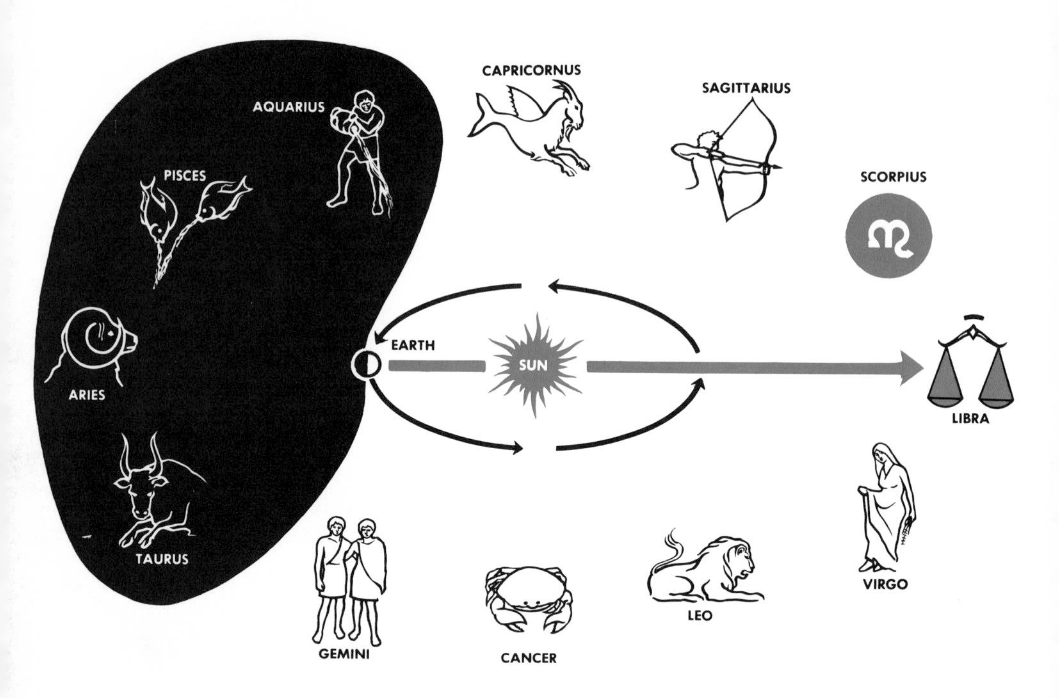


The sun is in the direction of Libra from October 29 to November 21. Scorpius is the astrological sign from October 24 to November 22.

Aquarius is in the evening sky along with Pegasus and Cepheus.

Lizard, Little Horse, Southern Fish, Crane, and Indian are near the meridian early in the evenings at mid-month.

M15, M39, and Delta Cephei are objects of interest as seen with binoculars.

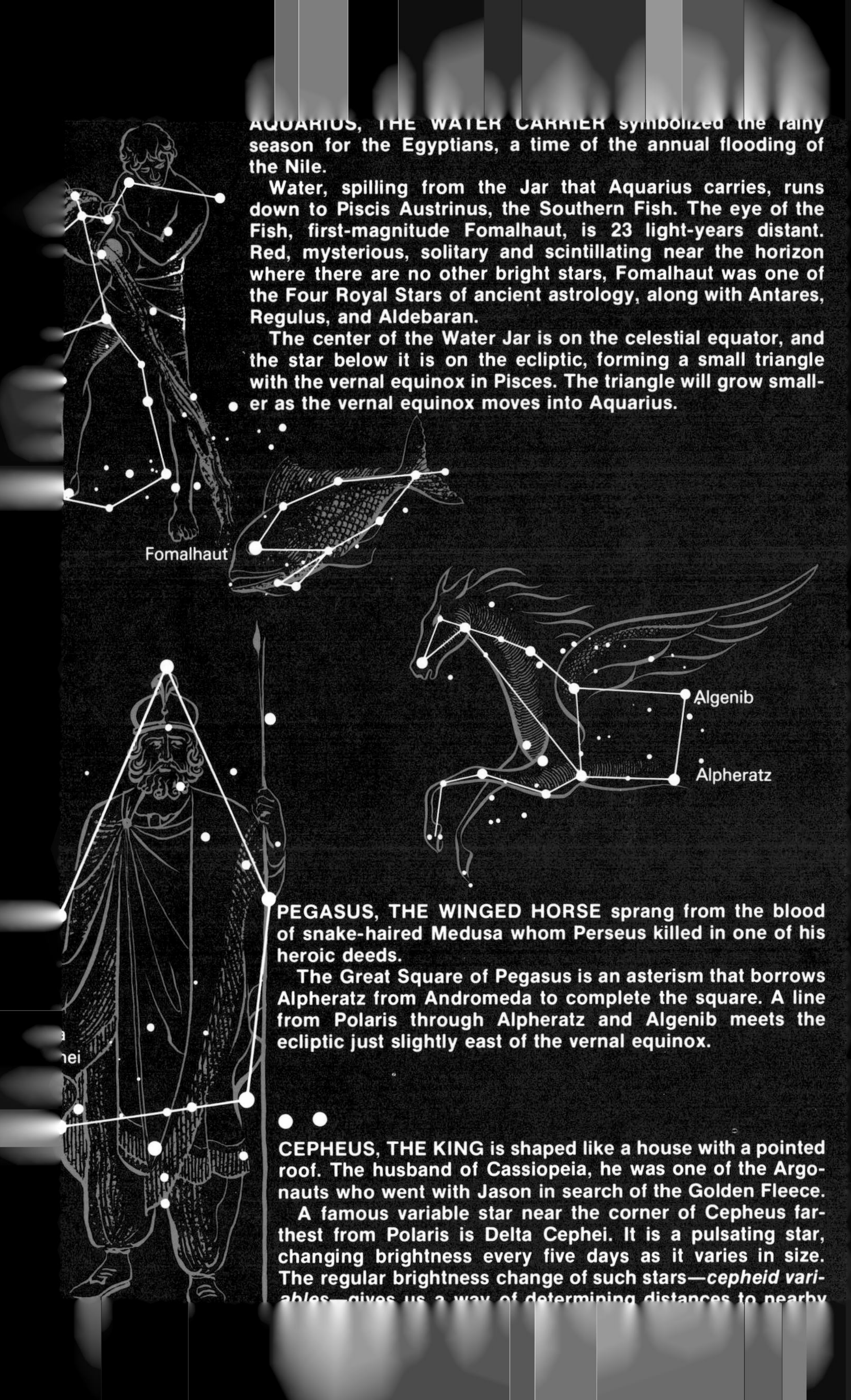


**AQUARIUS, THE WATER CARRIER** symbolized the rainy season for the Egyptians, a time of the annual flooding of the Nile.

Water, spilling from the Jar that Aquarius carries, runs down to Piscis Austrinus, the Southern Fish. The eye of the Fish, first-magnitude Fomalhaut, is 23 light-years distant. Red, mysterious, solitary and scintillating near the horizon where there are no other bright stars, Fomalhaut was one of the Four Royal Stars of ancient astrology, along with Antares, Regulus, and Aldebaran.

The center of the Water Jar is on the celestial equator, and the star below it is on the ecliptic, forming a small triangle with the vernal equinox in Pisces. The triangle will grow smaller as the vernal equinox moves into Aquarius.

**PEGASUS, THE WINGED HORSE** sprang from the blood of snake-haired Medusa whom Perseus killed in one of his heroic deeds.

The Great Square of Pegasus is an asterism that borrows Alpheratz from Andromeda to complete the square. A line from Polaris through Alpheratz and Algenib meets the ecliptic just slightly east of the vernal equinox.

**CEPHEUS, THE KING** is shaped like a house with a pointed roof. The husband of Cassiopeia, he was one of the Argonauts who went with Jason in search of the Golden Fleece.

A famous variable star near the corner of Cepheus farthest from Polaris is Delta Cephei. It is a pulsating star, changing brightness every five days as it varies in size. The regular brightness change of such stars—*cepheid variables*—gives us a way of determining distances to nearby

Pegasus prances to the meridian, his front feet on the Lizard and his head near the Little Horse. The Southern Fish swims in the stream of water pouring from Aquarius' Water Jar, and the Crane wades in the water near the southern horizon. The stars of Aquarius "altered the air and seasons in a wonderful, strange, and secret manner." And the Almanac of 1386 says of the sign, "It is gode to byg castellis, and to wed, and lat blode." The symbol for Aquarius ≈ shows undulating lines of waves.

A line curving from Polaris runs through Scheat and Markab in Pegasus, and continues down to Fomalhaut in PISCES AUSTRINUS, THE SOUTHERN FISH. Viewed from a latitude of 40° N, Fomalhaut is only 10 degrees above the horizon—the brightest star in the south at this time of year. *Fomalhaut* is Arabic for Fish's Mouth, but the star is more commonly referred to as the Eye of the Fish.

GRUS, THE CRANE wades in the River to the south of the Southern Fish. The high-flying Crane was a symbol of the star observer in ancient Egypt. Grus was also seen as a Flamingo "striking with his bill at the South Fish."

Between the Crane and PAVO, THE PEACOCK is another Bayer constellation—INDUS, THE INDIAN. It represents an American Indian, but he's much too far south to be seen from most of the United States.

Hevelius, picking up stars from Cygnus and Andromeda, created that "strange weasel-built creature with a curly tail," LACERTA, THE LIZARD. His alternative name, *Stellio,* is that of a small amphibious salamander with starlike dorsal spots. Three hundred years ago the faint star group was named to honor Louis XIV. A century later it was renamed to honor Frederick the Great. The Lizard remains.

A little group of 4th and 5th magnitude stars between the Horse and Dolphin is EQUULEUS, THE LITTLE HORSE. Hipparchus was supposed to have named the constellation, but somehow the name was lost. Hood dolefully wrote in 1590, "This constellation was named of almost no writer . . . and therefore no certain tail or historie is delivered thereof, by what means it came into heaven."

Between the two Horses is the M 15 globular cluster at a distance of 30,000 light-years. Near the heads of Lizard and Swan is a large open cluster of 6th magnitude stars, M39.

Capella
CAMELOPARDALIS
URSA MINOR
Kochab
Thuban
URSA MAJOR
Alkaid
CANES VENATICI
BOÖTES
Polaris
LITTLE DIPPER
PERSEUS
Mirfak
CASSIOPEIA
DRACO
CORONA BOREALIS
DOUBLE CLUSTER
M 103
Ruchbah
Algol VARIABLE
M 34
Navi
CEPHEUS
Alderamin
Eltanin
M 13
Schedar
Caph
Delta Cephei VARIABLE
Almach
Vega
HERCULES
TRIANGULUM
M 31
LACERTA
M 39
Deneb
LYRA
Beta Lyra
M 57
Mirach
ANDROMEDA
ARIES
Hamal
M 33
CYGNUS
Sheratan
Alpheratz
Scheat
Albireo
Ras-Algethi
VULPECULA
GREAT SQUARE
M 27
SAGITTA
Rasalhague
OPHIUCHUS
PEGASUS
PISCES
Algenib
Markab
DELPHINUS
Altair
AQUILA
Enif
M 15
EQUULEUS
SERPENS
CIRCLET
WATER JAR
March Equinox
M 11
SCUTUM
CETUS
AQUARIUS
Diphda
Ecliptic
CAPRICORNUS
Nunki
Fomalhaut
PISCIS AUSTRINUS
TEAPOT
SCULPTOR
SAGITTARIUS
MICROSCOPIUM
Ankaa
GRUS
CORONA AUSTRALIS
Al Nair
TELESCOPIUM
INDUS
PAVO
Achernar
TUCANA
October

# NOVEMBER

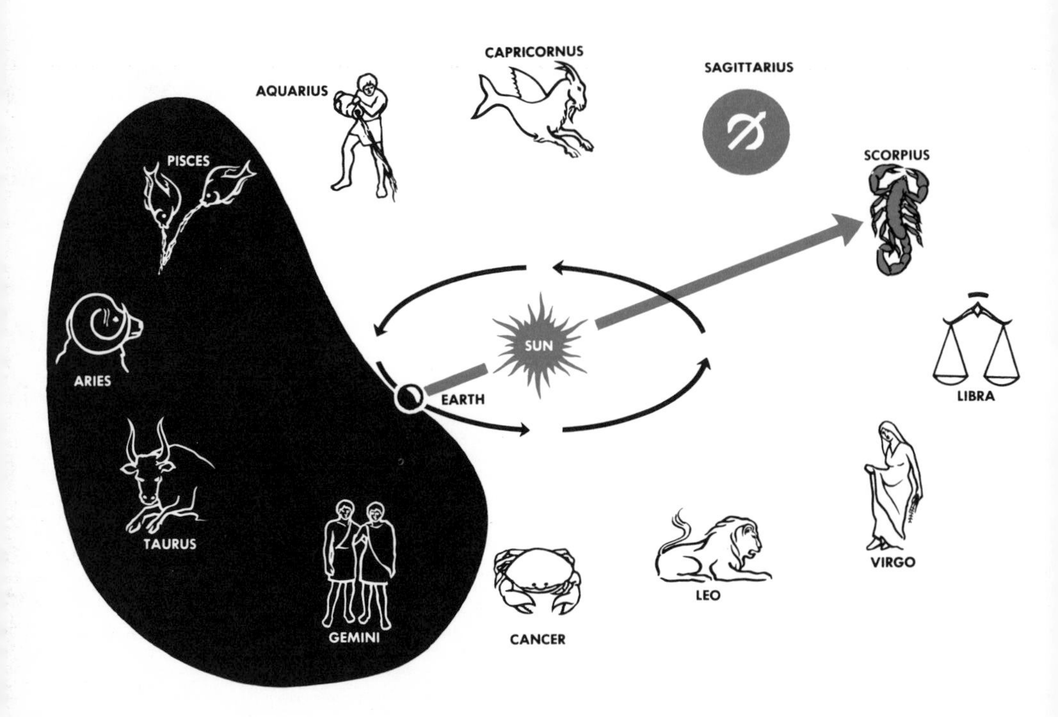


The sun is in the direction of Scorpius
from November 21 to November 29,
and in Ophiuchus from November 30 to December 16.
Sagittarius is the astrological sign
from November 23 to December 21.

Pisces is in the evening sky
along with Cassiopeia and Andromeda.

Phoenix, Tucana, and Sculptor
are near the meridian
early in the evenings at mid-month.

The Andromeda Galaxy, M31,
is a fascinating deep-space object.

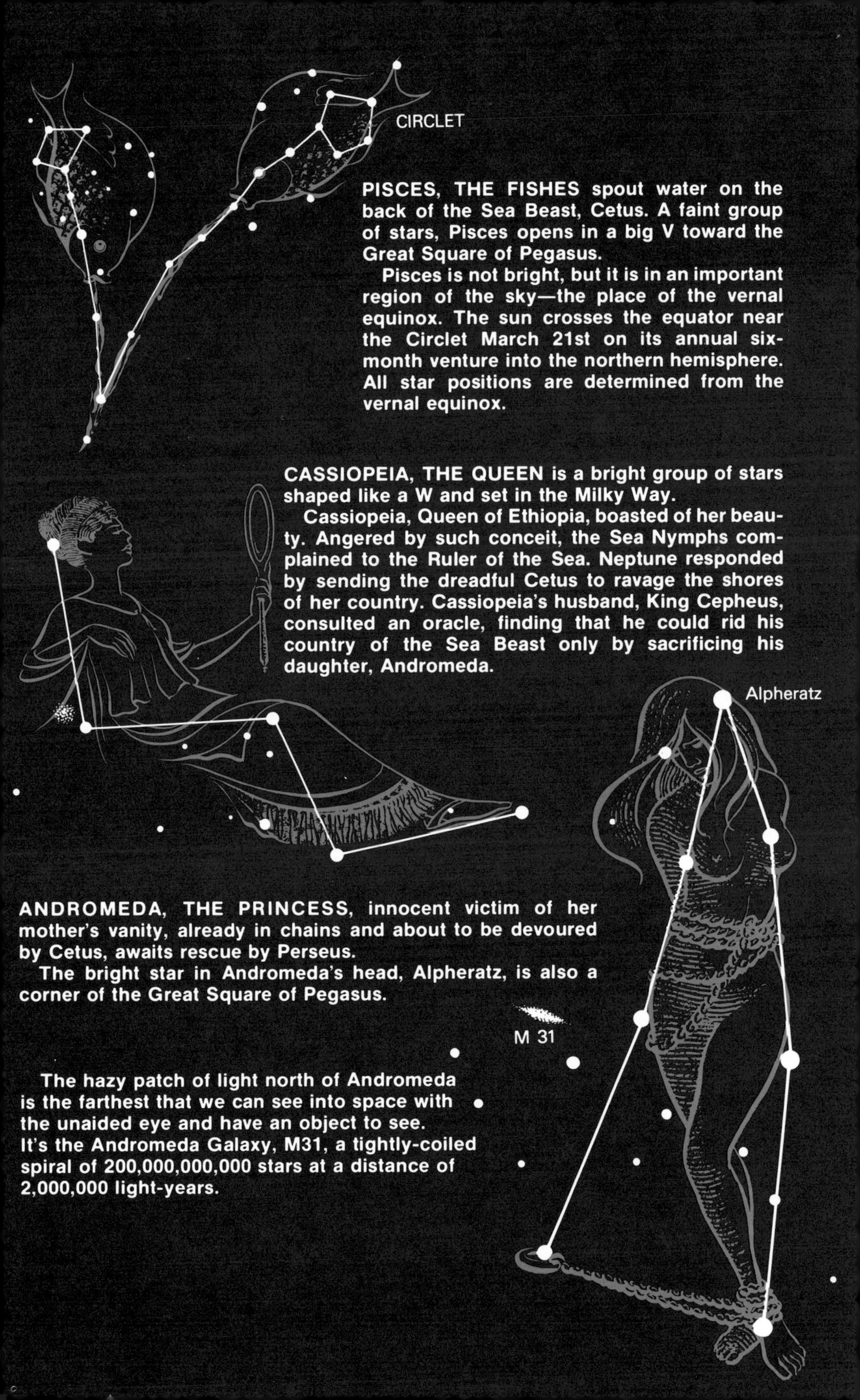


**PISCES, THE FISHES** spout water on the back of the Sea Beast, Cetus. A faint group of stars, Pisces opens in a big V toward the Great Square of Pegasus.

Pisces is not bright, but it is in an important region of the sky—the place of the vernal equinox. The sun crosses the equator near the Circlet March 21st on its annual six-month venture into the northern hemisphere. All star positions are determined from the vernal equinox.

**CASSIOPEIA, THE QUEEN** is a bright group of stars shaped like a W and set in the Milky Way.

Cassiopeia, Queen of Ethiopia, boasted of her beauty. Angered by such conceit, the Sea Nymphs complained to the Ruler of the Sea. Neptune responded by sending the dreadful Cetus to ravage the shores of her country. Cassiopeia's husband, King Cepheus, consulted an oracle, finding that he could rid his country of the Sea Beast only by sacrificing his daughter, Andromeda.

**ANDROMEDA, THE PRINCESS**, innocent victim of her mother's vanity, already in chains and about to be devoured by Cetus, awaits rescue by Perseus.

The bright star in Andromeda's head, Alpheratz, is also a corner of the Great Square of Pegasus.

The hazy patch of light north of Andromeda is the farthest that we can see into space with the unaided eye and have an object to see. It's the Andromeda Galaxy, M31, a tightly-coiled spiral of 200,000,000,000 stars at a distance of 2,000,000 light-years.

The Milky Way stretches east-west across the northern sky; Cygnus, Cepheus, Cassiopeia, Perseus, and Auriga lie within its boundaries. The watery constellations are sprawled across the southern skies, except for Pisces.

The two Fish in Pisces lie north of the ecliptic, widely separated from each other. The Fishes are now the Leaders of the Celestial Host, for here is the equinoctial point—the First Point of Aries. The Fishes, in classical Latin thought, were those that carried Venus and her son Cupid out of danger from Typhon. The Egyptians refrained from eating sea fish, and in their hieroglyphics used the fish to represent something odious.

A "dull, treacherous, phlegmatic sign," Pisces also "doth signifie weariness, sadness, poverty, deceit, feare, sorrow, blasphemy, ambushments, prisons and captivities."

PHOENIX, the beautiful bird of Egyptian mythology, lived in the Arabian desert for a half-thousand years or more. Then it consumed itself in fire, rising renewed from the ashes to start all over again. So it is a celestial symbol of a cyclic period. Its brightest star, Ankaa (93 light-years), is a solitary star almost directly south of the vernal equinox; and, viewed from 40° N, lies low on the horizon.

TUCANA, THE TOUCAN lies farther south, close to Achernar. The toucan is a brightly colored, fruit-eating bird, with down-curving beak—a native of South America. Most notable in the constellation is the great globular cluster, 47 Tucanae, with its rose-tinted center. It's a naked-eye cluster at a distance of 20,000 light-years, rivaling the Omega Centauri, Hercules, and Beehive clusters.

SCULPTOR, THE SCULPTOR lies between Cetus and Phoenix. La Caille called the constellation *l'Atelier du Sculpteur,* the Sculptor's Studio. It became the ponderous *Bildhauerwerkstatte* for the Germans but now is merely Sculptor. The south galactic pole lies in this constellation.

The Big Dipper is farthest to the north beyond Polaris; Cassiopeia at her most favorable southerly position,and Andromeda spans the United States, her head passing over New Orleans and her feet over Boston. The Great Spiral Galaxy, M31, passes over New York and Chicago, at a latitude of 41° N.

Caph, Schedar, Navi, and Ruchbah are the bright stars that are named in Cassiopeia; the westernmost star does not have a name. Cassiopeia sits in a rich region of many open clusters and gaseous nebulae. Fan-shaped M103 glows from 6,200 light-years.

URSA MAJOR
Kochab
URSA MINOR
LITTLE DIPPER
LYNX
Polaris
DRACO
HERCULES
CAMELOPARDALIS
Eltanin
Capella
M37
AURIGA
KIDS
M 38
Elnath
CEPHEUS
Alderamin
Vega
LYRA
Beta Lyra
M 57
DOUBLE CLUSTER
CASSIOPEIA
M103
Navi
Ruchbah
Delta Cephei
VARIABLE
Caph
Schedar
Deneb
Mirfak
PERSEUS
M39
CYGNUS
Albireo
Algol
VARIABLE
M34
Almach
M31
LACERTA
VULPECULA
M45
PLEIADES
TRIANGULUM
ANDROMEDA
M27
SAGITTA
Mirach
Ecliptic
HYADES
M33
Scheat
ARIES
Hamal
Alpheratz
Altair
Sheratan
GREAT SQUARE
PEGASUS
DELPHINUS
M 15
AQUILA
Enif
Algenib
Markab
EQUULEUS
Menkar
PISCES
WATER JAR
CIRCLET
Mira
VARIABLE
March Equinox
CETUS
Ecliptic
ERIDANUS
AQUARIUS
CAPRICORNUS
Diphda
Fomalhaut
PISCIS AUSTRINUS
MICROSCOPIUM
FORNAX
SCULPTOR
Ankaa
PHOENIX
GRUS
Acamar
ERIDANUS
Al Nair
HOROLOGIUM
INDUS
Achernar
TUCANA
PAVO
HYDRUS
RETICULUM
November

# DECEMBER

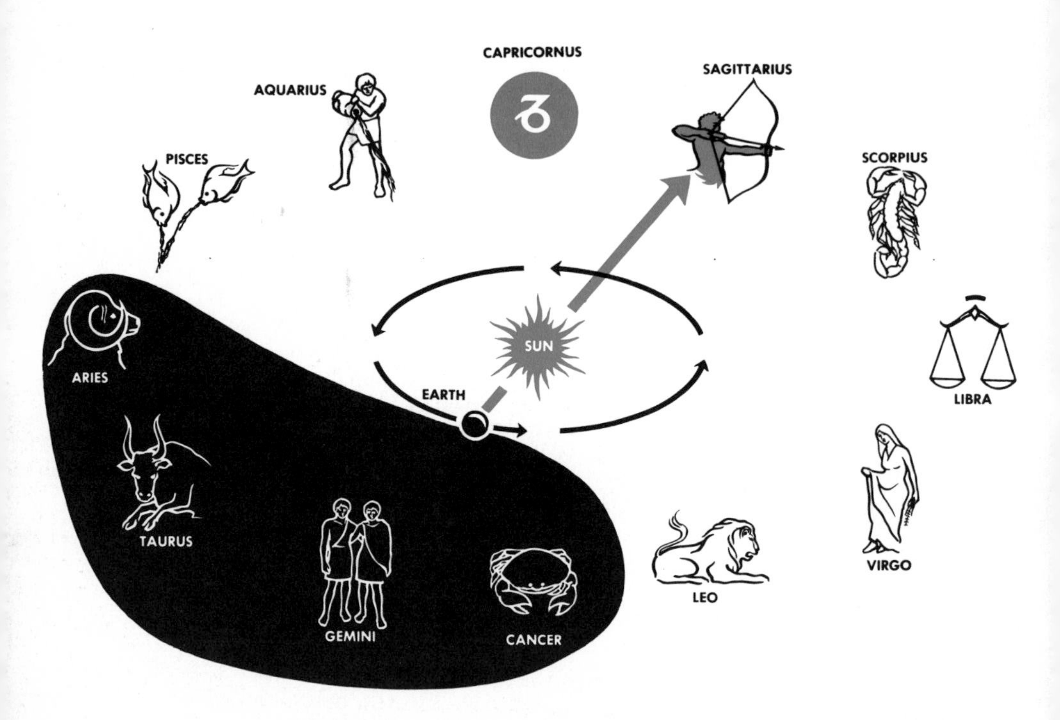


The sun is in the direction of Sagittarius
from December 16 to January 18.
Capricornus is the astrological sign
from December 22 to January 19.

Aries is in the evening sky
along with Perseus and Cetus.

Water Monster, Triangle, Clock,
and Chemist's Furnace
are near the meridian
early in the evenings at mid-month.

M 33, M 34, and the variable star Algol
are objects of interest in binoculars.

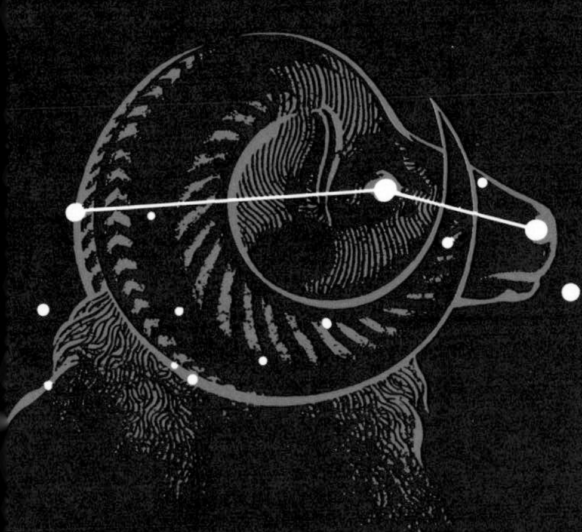

**ARIES, THE RAM** is a small group of stars representing the animal with the Golden Fleece. Mercury gave Aries to Queen Nephele of Thessaly when she feared that harm might befall her children. She put them on the Ram, sending them to a safe place. But as they flew across the sea between Europe and Asia, the girl, Helle, fell off at a place known today as the Hellespont.

The Boy, Phrixus, arrived safely in Colchis. Here the ram was sacrificed to Jupiter, and his fleece placed in a sacred grove guarded by a dragon. High adventure was the lot of Jason and the Argonauts who performed dangerous deeds to take the Golden Fleece back home.

**PERSEUS, THE HERO** is a sharply pointed arrow-shaped group of stars set in the Milky Way, pointing toward Cassiopeia. He holds the Medusa's head in one hand and he is about to dispatch Cetus and rescue Andromeda.

The winking Demon Star, Algol, was thought by the ancients to be the most violent, unfortunate, and dangerous object in the sky, for it changed brightness in the unchanging and eternal heavens. Algol is an eclipsing pair—a *binary system* bound together by gravitation. The smaller of the two is the brighter. Every three days the bigger, dimmer star moves in front of the smaller one, reducing its brilliance to a third for 9½ hours.

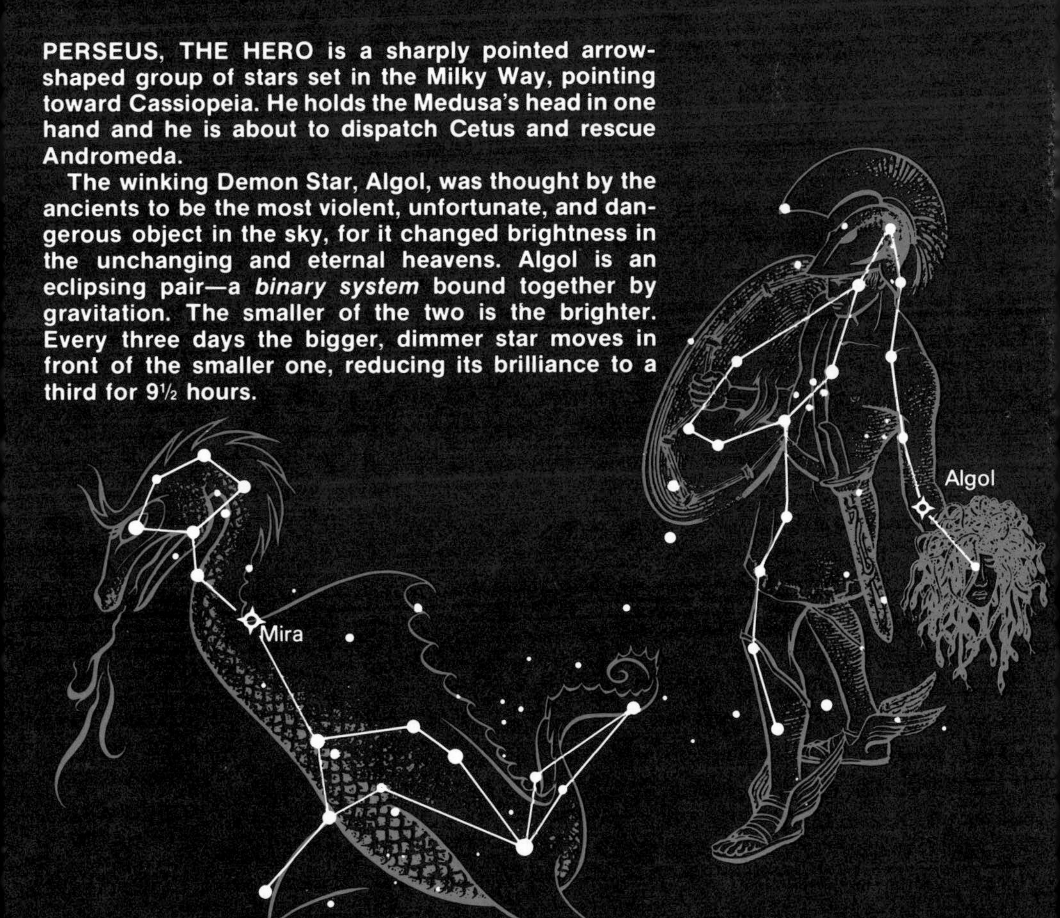


**CETUS, THE SEA BEAST** is the monster Perseus slew to rescue Andromeda. Mysterious Mira, The Wonderful, is almost as bright as Polaris for a month or two a year, then disappears completely from sight. It is a long-period variable, 300 times the sun's diameter—a huge pulsating star varying in temperature as well as size.

Draco, the Dragon moves behind Polaris, ceding his place in the sky to the two monsters, Cetus and Hydra. Perseus and Andromeda are both sharply-pointed star groups, and the Hero lies at the feet of the Princess.

The "Ram that bore unsafely the burden of Helle" is in size a sharp contrast to the sprawling Fishes to its west. Aries was a dreaded sign among the ancient astrologers, indicating passionate temper and bodily hurt. Seventeenth-century almanacs gloomily predicted that "many shall die of the rope" when the sun is in that sign. The almanac of 1592 declares that "the infant born in Aries disposeth his life in that signe, and also placeth in the sayde signe his speech, wisedome, augmentation of all his workes, his beginning, his name and the originall of his life and yeares."

A line curving up from the Pleiades joins another curving line of stars at a sharp point—the head of Perseus. The Double Cluster in Perseus lies at a distance of 4,300 light-years. M34 in Perseus is a cluster of 6th magnitude stars just visible to the unaided eye. It is 1,300 light-years from earth. Near Mirfak is a pulsar sending out bursts of radiation 85 times a minute.

TRIANGULUM, THE TRIANGLE is an old constellation more noted by the ancients than by us, for it represented Sicily, reproduced in the sky by Jove upon Ceres' pleading. The first minor planet to be discovered—Ceres—was found here on January 1, 1800.

The sharp point of the Triangle lies midway between Hamal and Mirach, and M 33 is only slightly west of that point. Magnitude eight, M 33 is hard to find because of its size and poor contrast with the background. A large telescope photographs it as a spiral galaxy, 2,400,000 light-years distant.

FORNAX and HOROLOGIUM lie on opposite banks of the celestial river Eridanus. Fornax was called *Fornax Chemica,* the Chemist's Furnace, by La Caille, in 1769. A few years later, Bode changed it to *Apparatus Chemicus,* honoring the celebrated chemist Antoine Lavoisier. Now it is simply Fornax. *Horologium Oscillatorium,* the Pendulum Clock, has been abbreviated to Horologium.

HYDRUS, THE WATER MONSTER lies in the extreme southern part of the sky, his tail near Achernar and his head near the south celestial pole. A yellow 3rd magnitude star in Hydrus is the nearest bright star to the south celestial pole—but it is still 12 degrees from it.

DRACO
Kochab
LITTLE DIPPER
HERCULES
Eltanin
URSA MINOR
M81
DRACO
URSA MAJOR
Vega
Beta Lyra
LYRA
M57
Polaris
LYNX
CAMELOPARDALIS
CEPHEUS
Alderamin
Deneb
CYGNUS
Delta Cephei
VARIABLE
CASSIOPEIA
M39
Castor
DOUBLE CLUSTER
M103
Navi
Caph
Ruchbah
Schedar
VULPECULA
Capella
Mirfak
KIDS
PERSEUS
LACERTA
M 37
AURIGA
M35
M38
M34
M31
Algol
VARIABLE
Almach
ANDROMEDA
Elnath
June Solstice
Mirach
M33
Scheat
PEGASUS
Alpheratz
M1
TRIANGULUM
M45 PLEIADES
GREAT SQUARE
Enif
TAURUS
Hamal
ARIES
Sheratan
Aldebaran
Markab
HYADES
Algenib
PISCES
ORION
Bellatrix
WATER JAR
CIRCLET
Alnilam
Mintaka
Menkar
Ecliptic
March Equinox
M42
Mira
VARIABLE
Rigel
CETUS
AQUARIUS
LEPUS
Diphda
FORNAX
Fomalhaut
ERIDANUS
SCULPTOR
PISCIS AUSTRINUS
Acamar
Ankaa
PHOENIX
CAELUM
GRUS
HOROLOGIUM
DORADO
Achernar
RETICULUM
HYDRUS
TUCANA
Large Magellanic Cloud
December

# HOW BIG? HOW FAR?

*Thirty Earths*
*to Tycho's crater;*
*a hundred Earths—and more—*
*to span the face of Sun*
*a hundred Suns away.*

*A hundred billion*
*spinning suns*
*wound in a whirling spiral*
*a hundred thousand*
*light-years across.*

*Two million years*
*to Andromeda's glow;*
*and a billion*
*island-universes beyond,*
*faint hints*
*of primordial light.*

The astronomer expresses stellar sizes, distances, and aggregations in numbers truly astronomical. Comparisons help us to visualize relationships that we but little comprehend. In this chapter, we will deal with the grand picture of the universe that magnitude gives, leaving precise values to astronomical tables.

## EARTH AS MEASURE

Earth-diameter is a convenient unit of measure in dealing with the relationship of matter and space near the Sun. Taking the Earth's 12,800-kilometer diameter as 1, then we find that the four planets closest to the Sun are of somewhat comparable size: Mercury's diameter is 0.4, Venus' 1.0, and Mars' 0.5.

The outer planets, except for Pluto, are much larger than the inner ones. Jupiter is 11 times the Earth's diameter, Saturn 10 times, Uranus and Neptune, 4; but Pluto's estimated diameter is less than the Earth's.

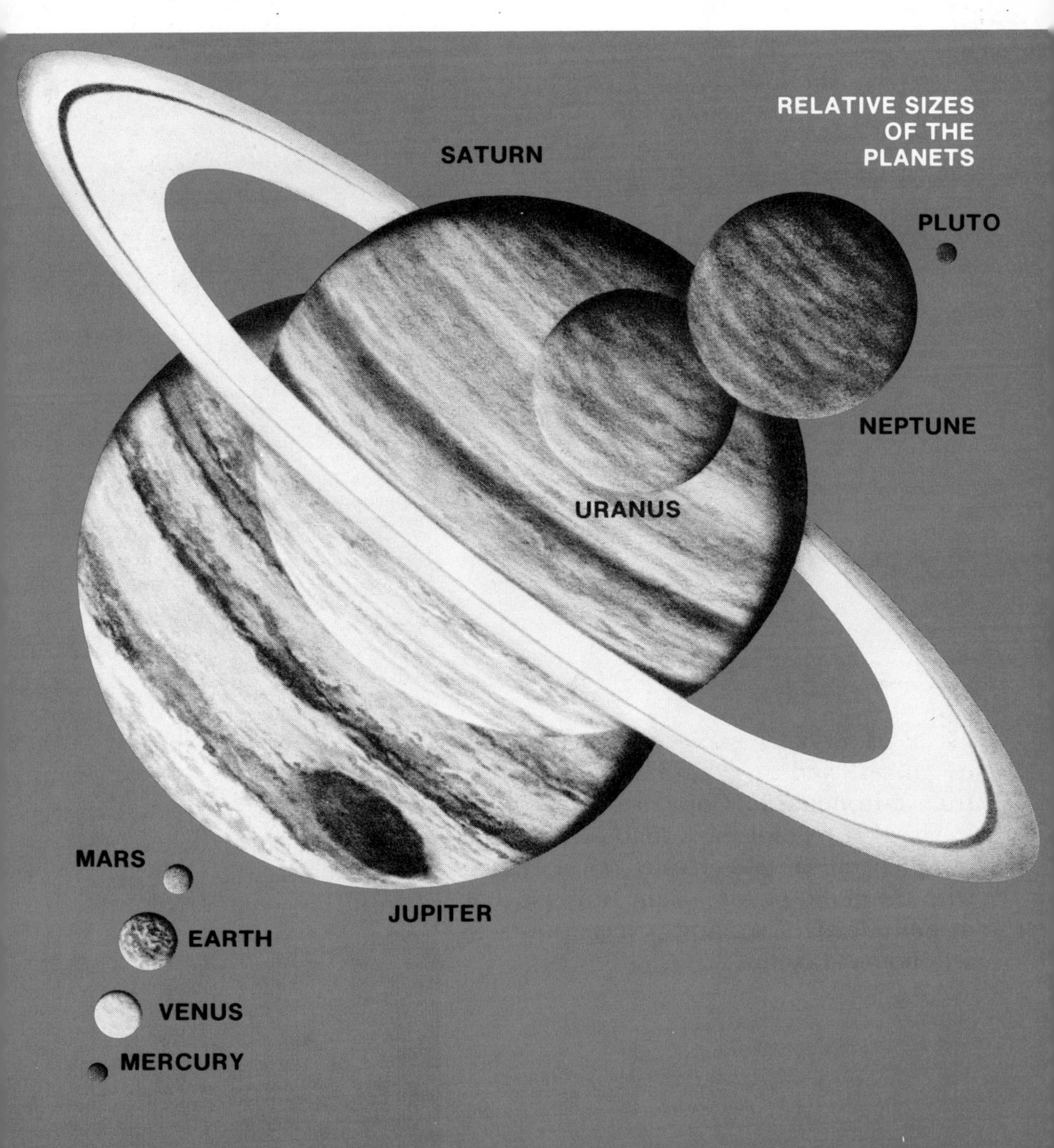

Circling the Earth is one of the largest planetary satellites in the solar system. Jupiter, Saturn, and Neptune have satellites larger than our Moon, but those moons are dwarfed by their enormous planets. Viewed from afar, the Earth-Moon system might well be called Twin Planets, not only for their nearness in size but also for their proximity in space.

The Moon *is* close—a hundred times closer than either Mars or Venus even at their nearest approaches. Light takes a second-and-a-quarter to travel the distance. Astronauts take three days. Thirty Earths can fit in the space between Earth and Moon. *Ratio, 30:1.*

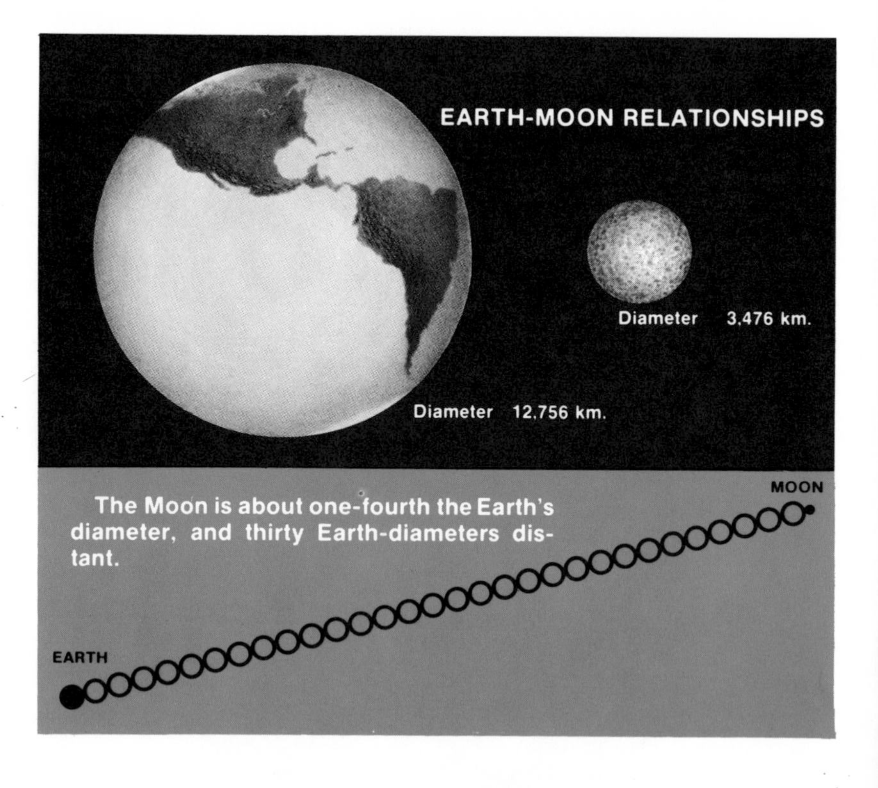


The Earth-diameter unit also gives us an idea of the size of the Sun. A hundred Earths can fit across its face with plenty of room to spare, for the Sun is nearly a million miles in diameter. Planet Earth is, on this scale, just a tiny speck compared to the star that energizes it. *Ratio, 100:1.*

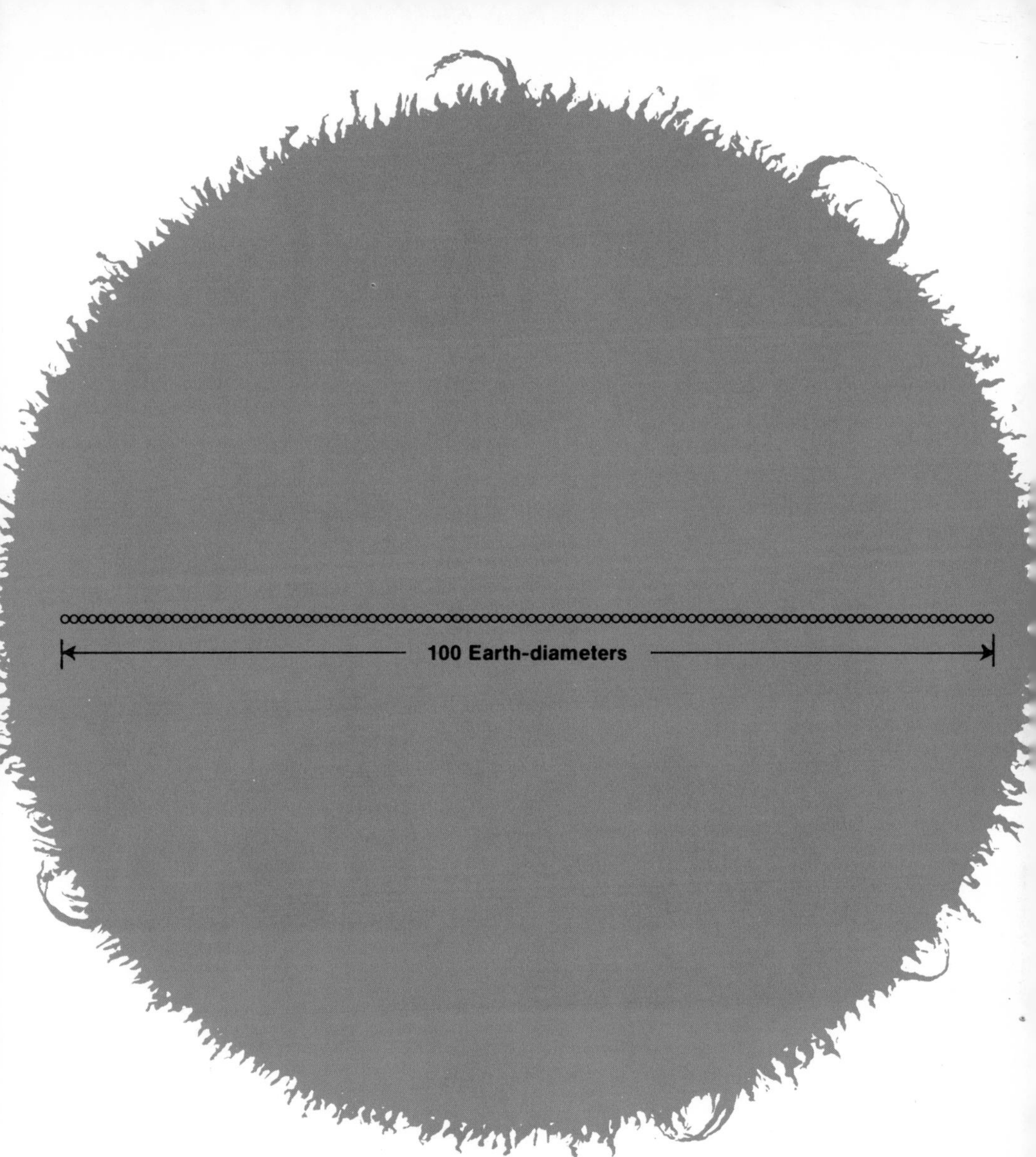


**The Sun's diameter is more than 100 Earths.**

But volume is something else again. Here the Earth shrinks to pinpoint significance. Since the volume of a sphere is proportional to the cube of its radius, then the Sun could contain 100 x 100 x 100 —a million Earths! *Ratio, 1,000,000:1.*

Two astronomical coincidences help to sharpen the picture. One is the Sun-Moon relationship; the other, the Sun-Earth distance.

## SUN–MOON RELATIONSHIP

The Sun is 400 times the diameter of the Moon, and 400 times more distant. Each subtends an angle of one-half degree in the sky. Occasionally the Moon moves in front of the Sun, just covering the solar disk, darkening the sky. Such a solar eclipse reveals the Sun's corona for a few minutes—seven at most. Without that coincidence, stellar theory would not be as advanced as it is today and we'd know correspondingly less about the other stars in the universe. *Ratio, 400:1.*

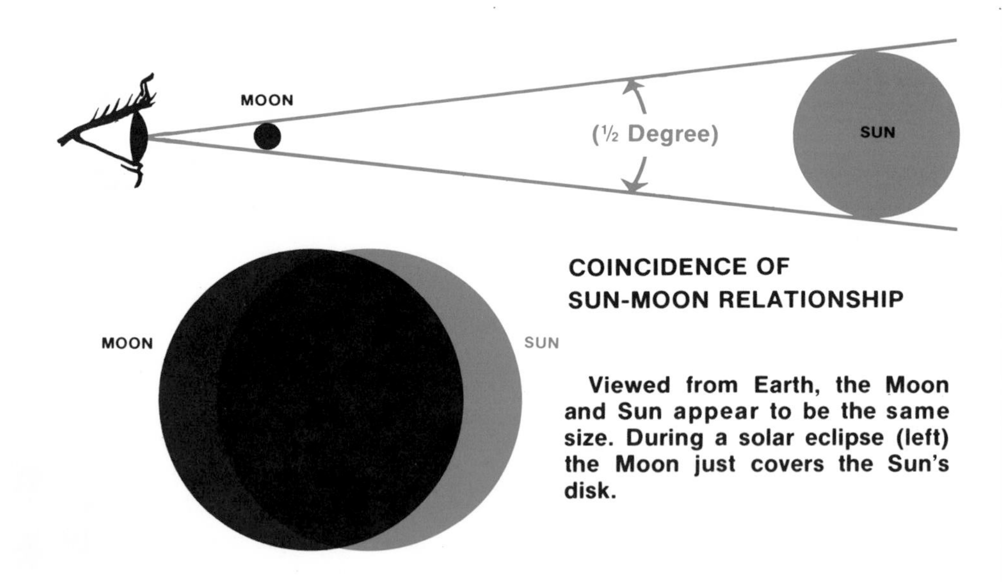


**COINCIDENCE OF SUN-MOON RELATIONSHIP**

**Viewed from Earth, the Moon and Sun appear to be the same size. During a solar eclipse (left) the Moon just covers the Sun's disk.**

## SUN–EARTH DISTANCE: THE ASTRONOMICAL UNIT

A hundred Earths can, as we have seen, fit across the Sun. Another hundred-to-one ratio is the coincidence of Sun-Earth distance: a hundred Suns can fit in the space between the Sun and Earth.

Earth-diameter has its limitations as a unit of measure; so does Sun-diameter. The most useful unit for measuring distances within the solar system is the Sun-Earth distance, or astronomical unit (AU). Taking the AU (149,600,000 km) as 1, the planetary distances stand in neat ratios that are easy to remember:

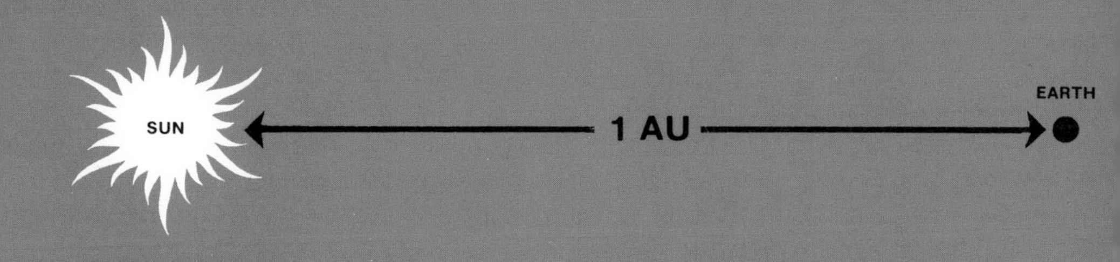


| | Planet | Distance from Sun (AU) |
|---|---|---|
| ☿ | Mercury | 0.4 |
| ♀ | Venus | 0.7 |
| ⊕ | Earth | 1.0 |
| ♂ | Mars | 1.5 |
| ♃ | Jupiter | 5 |
| ♄ | Saturn | 10 |
| ⛢ | Uranus | 20 |
| ♆ | Neptune | 30 |
| ♇ | Pluto | 40 |

Jupiter is five times as far from the Sun as is Earth. Saturn is ten times as far—a billion miles. Pluto, 40 times as far: 4,000 Sun-diameters.

Let the width of this page represent the AU: then Pluto is 40 pages distant. The Sun as seen from Pluto would be small and brilliant—about the size of Jupiter as seen from Earth, but far brighter than the full Moon. Since light intensity varies inversely as the square of the distance, the Sun is

$1/(40)^2$, or a sixteen-hundredth as bright at Pluto as it is at Earth.

Pluto, too, is a captive of the Sun, held in its path by the gravity of the central star. It is hard to see how a pinhead of matter (the Sun, on our scale) can influence—much less hold captive—the tiniest speck of dust, ⅛th the Earth's volume, at a distance of 40 pages. Yet, the Sun's gravitational influence extends far beyond this outermost planet as revealed in the long elliptical paths of comets, some of which may take thousands or even millions of years to return.

## THE MATTER NEAR THE SUN

Before leaving the solar system and the astronomical unit for a greater order of magnitude, let's take a look at the matter near the Sun as revealed in optical telescopes, as probed by radar telescopes, as sensed by infrared and ultraviolet detectors, as looked at by orbiting spacecraft, and as sampled by landing vehicles.

### Earth—the Blue Planet

We'll start with the Earth, for it is the model of the *terrestrial planets*—those planets of high density near the Sun. Densities of the terrestrial planets vary from Mars' 4.0 to the Earth's 5.5. The Earth has a nickel-iron core, but it is questionable whether Mars has or not. The Moon (density 3.3) is apparently without a metal core.

The large planets of low density, far from the Sun, including Jupiter, Saturn, Uranus, and Neptune, are the *Jovian* planets. Pluto's physical features are only estimated, not measured, for it is too far away to be seen clearly—like trying to locate a bowling ball at a distance of 1 km in complete darkness with only a flashlight. It is estimated to be about the size and mass of Mars.

Satellites have mapped the Earth's magnetic field far out into space. The field is not symmetrical as is that of a small magnet, but is flattened on the sunny side, where a shock wave is produced by the solar wind, and elongated on the dark, lee side. Electrically charged particles shot out by the Sun, particularly at times of high solar activity, are caught in the magnetic field and spiral in helixes from one geomagnetic pole to the other. One to two hundred km above the Earth's surface, the particles collide with the tenuous atmosphere. Light is given off, brightening the polar skies with the Aurora Borealis in the northern hemisphere and the Aurora Australis in the southern.

The Earth, traveling 30 times as fast as a bullet as it circles the Sun, runs head-on into bits of nickel-iron or stone. Friction with the atmosphere heats the particles into incandescence. Then they glow as "shooting stars," or *meteors*. Each day 25,000,000 meteors

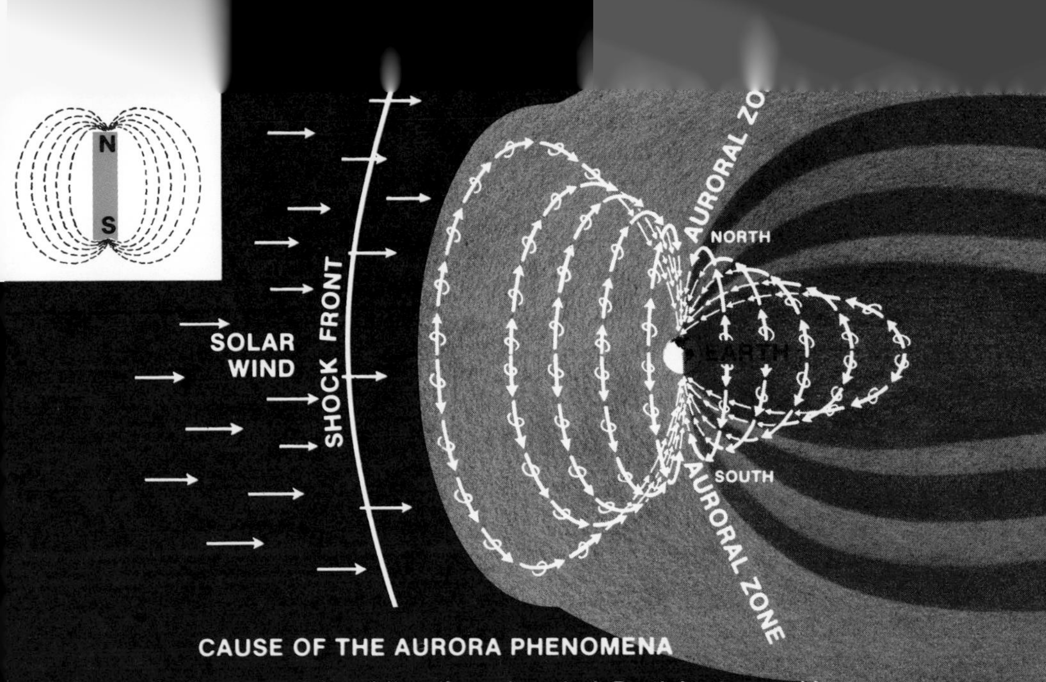


CAUSE OF THE AURORA PHENOMENA

Arrows indicate particles riding the solar wind. Particles trapped by Earth's magnetic field are constrained to magnetic lines of force, analogous to the arrangement of iron filings in the field of an ordinary bar magnet (inset). Converging on Earth's geomagnetic poles, their collision with the atmosphere produces the Aurora.

strike the Earth's atmosphere. Most of them are no larger than grains of sand, and burn up in a brief flash. Basketball-sized particles occasionally enter the atmosphere, traveling 15 to 75 km/second. Such large objects, *bolides,* turn night into day for a short time. Once in a while they may be bright enough to be seen in the daytime. Those which survive to strike the Earth itself are *meteorites.*

## Fleet-footed Mercury

Mercury is traveling 60 times faster than a bullet as it circles the Sun. Closest to the Sun of all planets, Mercury receives six times as much heat and light as does the Earth.

We must look toward the Sun to see Mercury. It is always within 30 degrees of the Sun, first on one side, then on the other. Polynesians called it *Na-holo-holo,* running to and fro. Visible only in the morning or evening twilight near the horizon, Mercury has been a difficult planet to study. But photos taken by Mariner spacecraft show its surface to be much like that of the Moon and Mars—pock-marked and covered with impact craters.

Because of its proximity to the Sun (0.4 AU), Mercury was once thought to be locked in synchronous rotation with the Sun, always keeping one face toward it, just as the Moon does toward the Earth. But radar studies of the planet show it turning on its axis in two months while traveling around the Sun in three months. It's a long, slow, hot day on Mercury.

## Exotic Venus

Venus, like Mercury, lies within the Earth's orbit, and appears in morning or evening twilight. Farther from the Sun than Mercury, Venus moves more slowly and remains in the evening sky for a longer period of time. Venus' *greatest elongation* (the farthest it gets from the Sun in angular distance as seen from Earth) is nearly 50 degrees. Then it sets three hours after sunset.

Venus lingers many months as an "evening star," going through phases as it comes from behind the Sun. The phasing is visible in a small telescope. Maximum brilliance is achieved after the planet passes greatest elongation and begins catching up to the Sun. When less than half of the planet is lighted as seen from Earth, it is so brilliant that it can be seen in the daytime. On dark nights it casts a shadow. Only the Sun and Moon outshine Venus.

Cloud-covered Venus was once thought to have a cool surface beneath its 50-km atmospheric blanket. But it's hotter than a kitchen oven—600°C (900°F)—hot enough to melt lead. The heat is due to the "greenhouse effect." Sunlight striking the planet's surface is reradiated at longer wavelengths. Heat builds up as the infrared radiation is trapped by the carbon dioxide atmosphere.

Atmospheric pressure on Venus is 100 times that on Earth, or about the same as that beneath 1,300 meters of ocean water. The severe conditions of temperature and pressure are similar to those in power plants on Earth where high-pressure steam is used to turn turbines to produce electricity.

The day on Venus is longer than its year. And it spins backward, rotating slowly retrograde in 243 days as it travels around the Sun in 228 days. Measurements show little or no magnetic field, possibly a consequence of its slow spin rate.

## ♂ Fiery-red Mars

Earth is just right for life, as we know it, to exist. Venus is too hot and cloudy. Mars is too clear and cold.

Mars is a twin of the Earth in many respects. The inclination of its equator to its orbit around the sun is only a half-degree greater than the Earth's. The Martian day is 37 minutes longer than the Earth-day. Mars, though, is only half the diameter, and 1/10th the mass of Earth. Small size means that it has not been able to store much heat.

The atmosphere on Mars is thin—1/100th that of Earth. Most of the time we can see its red surface, but dust storms sometimes fill its atmosphere with red clouds, as powerful winds transfer heat from one region to another. Mars is cold at the poles, warm at the equator. Its equatorial region warms up to about the same temperature as the Earth's tropic zone during the day, but the heat radiates away at night and the temperature drops to nearly -75°C.

Fifty years ago, it was supposed by some that Mars might have a civilization more advanced than ours. Great canals, that seemed to be visible in large telescopes, were thought to have brought wa-

**Mars has always fascinated Earth-bound man. Drawings such as this, indicating "canals" and "vegetation," were the best maps of the planet until the advent of photography by spacecraft. North is at the top.**

ter from polar ice caps to the parched equatorial region where most of the Martians lived. Now it is known that the "canals" were the result of human imagination working beyond the limit of resolution of optical telescopes. Mars is dry like the Moon. Impact craters are scattered all over its surface. And the biggest volcano in the solar system—tens of times bigger than any on Earth—is on a planet half the size of Earth. The red planet has chaotic regions of deep canyons and fault-block mountains—a topography known nowhere else in the solar system. Erosional patterns seem to indicate that water once flowed on Mars.

Mars fluctuates greatly in brightness. Maximum brilliance is reached when it comes to within 55 million km of Earth; but its luster fades as it swings around the far side of the Sun. Every two years Earth passes Mars at *opposition;* then we see it best in the midnight sky, opposite the direction of the Sun.

## ♃ Jupiter—the Giant

The biggest planet in the solar system is 318 times more massive than Earth, 2½ times more massive than all the other planets put together. Yet it is tiny compared to the Sun, having only a thousandth the solar mass. Jupiter is a borderline case between a planet and a star, for it is nearly the maximum size that a body of "cold" hydrogen gas can be. More massive concentrations of hydrogen squeeze into smaller volumes; considerably more massive bodies form stars. Despite its great mass, Jupiter's density is 1.34—only slightly greater than that of water.

**Jupiter, showing the mysterious Great Red Spot in the southern hemisphere. Satellite at right casts an eclipse shadow on the face of the planet.**

Jupiter emits strong signals at radio frequencies. Radio telescopes find that it is second only to the Sun as a strong radio source. It is also a strong emitter of infrared. Measurements indicate 2½ times as much infrared energy coming from the planet as it receives from the Sun, indicating a source of heat beneath its clouds.

Jupiter reflects 45% of the sunlight striking its clouds, making it second only to Venus in brightness. Rapid rotation (Jupiter spins in less than half the time of Earth, yet its diameter is 11 times as great) stretches its clouds out into bands. Irregular spots interrupt the bands for various lengths of time, from two hours to two years. The Great Red Spot, three earth-diameters in length, was revealed by Pioneers 10 and 11 to be a special kind of storm peculiar to Jupiter. A century ago the Red Spot was definitely red, but now it is no longer highly colored.

Jupiter has a strong magnetic field. The orbital position of its satellite Io triggers strong bursts of radio energy. Pioneer 10 sailed through the radiation belts of Jupiter relatively unscathed. Pioneer 11, though, did not fare so well, receiving shocking electrical jolts that nearly put it out of commission.

The continually changing position of Jupiter's satellites, the Red Spot, the cloud banding, and the polar flattening make the largest planet a telescopic favorite. Four of its moons are easily seen in small telescopes. Five times as far from the Sun as is Earth, Jupiter moves slowly through the heavens, spending an average of one year in each zodiacal constellation.

## ♄ Saturn—the Ringed

**Saturn is light enough to float in water.**

Cloud-covered Saturn is twice as far from the Sun as is Jupiter. It spends 2½ years in each zodiacal constellation during its 30-year trip around the Sun. Lowest in density of all the planets (0.7), it could actually float on water with three-tenths of its mass rising above the surface.

The rings of Saturn are tilted 27 degrees to its plane of orbit around the Sun. First one polar region, then the other, is presented to our view. Between these extreme positions, we see the rings edge-on; at that time they disappear in all but the largest telescopes, for they're so thin. The rings, perhaps 20 km in thickness, are made of billions of particles of ice, ranging in size from marbles to basketballs, and each orbiting the planet at its own speed and distance. Saturn, like Jupiter, emits more energy than it receives from the Sun, suggesting an internal source of heat.

### Uranus—the Roller

William Herschel discovered Uranus in 1781. A dark, clear night reveals it to the unaided eye at opposition as a faint green object. Twice as far from the Sun as Saturn, Uranus takes 84 years to make the solar circuit, spending an average of 7 years in each zodiacal group. Uranus is a strange planet, lying on its side, rolling around the Sun with its axis in nearly the same plane as the ecliptic. Its five satellites are visible only in large telescopes.

### Neptune—the Predicted

But Uranus was not behaving according to Newtonian theory. French and English astronomers suspected the presence of a still more distant planet tugging on it, causing it alternately to speed up and slow down. The new planet (now Neptune) was predicted to be in Aquarius. A lengthy search led to its discovery in 1846 near the predicted place, further validating Newtonian physics. Neptune's atmosphere contains methane and hydrogen, and it appears as a small, greenish disk in a telescope. Neptune spends an average of 14 years in each zodiacal constellation.

### Pluto—the Plodder

Pluto, too, was predicted. But its discovery was an accident, for its calculated position in Gemini, though mathematically correct, was based on invalid data. Fortunately, it was near the predicted position. Finding it was difficult, however, for it was located in the Milky Way, and each photographic plate showed a third of a million stars. A 20-year search by Lowell Observatory astronomers finally revealed Pluto plodding along at 4.4 km/sec in the cold darkness some 4 billion miles from the Sun.

But the most distant planet is not always the farthest from the Sun. Pluto's eccentric path will bring it within the orbit of Neptune between 1979 and 1998. No collision is possible, though, because of the high inclination of Pluto's orbit to the ecliptic.

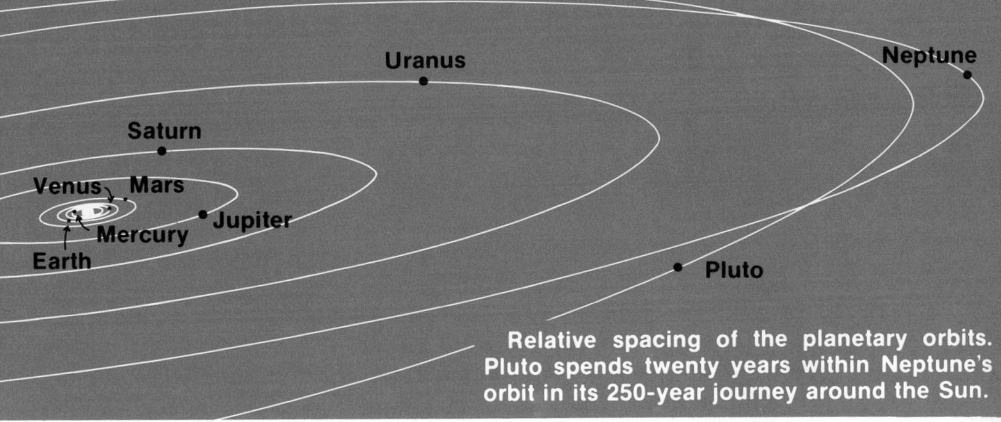

Relative spacing of the planetary orbits. Pluto spends twenty years within Neptune's orbit in its 250-year journey around the Sun.

## FROM ASTRONOMICAL UNIT TO LIGHT-YEAR

Pluto stands at the edge of the solar system, 40 AU from the Sun. The next star is 7,000 times farther. The AU is too small a unit to be useful in measuring great stellar distances. Here it gives way to the *light-year,* a more convenient unit. The light-year is a measure of distance, not time—the distance that light travels in one year: about 9,500,000,000,000 km.

We ordinarily consider light as being instantaneous, at least when we're dealing with objects at the Earth's surface. At a speed of 300,000 km/second, light can travel seven times around the world in one second—faster than we can count aloud or make seven circles on a piece of paper.

But the travel time of light is an important factor in astronomical distances. Light from the Sun reaches Earth in 8 minutes, Jupiter in 45 minutes, and Saturn in an hour and a half. It passes Pluto in a quarter of a day and speeds on into space. Then it travels for days . . . weeks . . . months . . . years . . . slightly more than four years, to reach the nearest star, Alpha Centauri, 40,850,000,000,000 km distant.

The most distant stars that we can see with the unaided eye are a thousand times farther than Alpha Centauri.

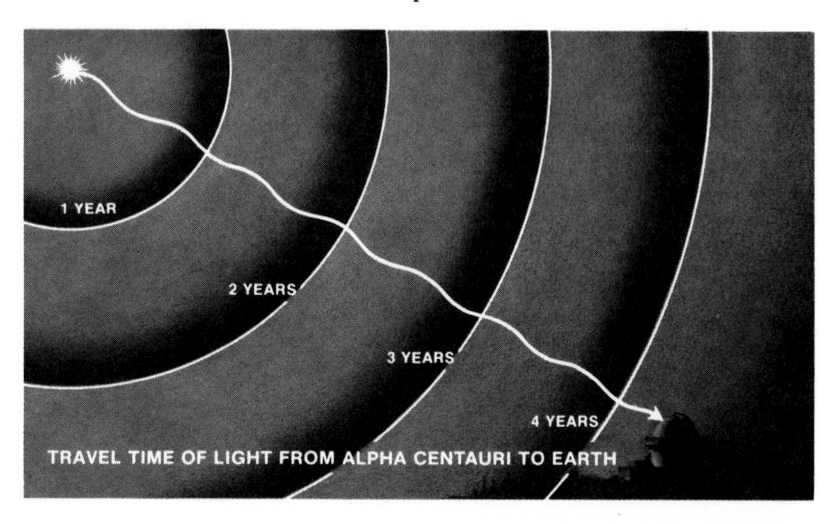


TRAVEL TIME OF LIGHT FROM ALPHA CENTAURI TO EARTH

## STELLAR SIZE AND EVOLUTION

The Sun is nearly a million miles in diameter. The biggest red giant is a thousand times larger; the smallest luminous star is a hundred thousand times smaller. Since stellar size, mass, and longevity are interrelated, the amount of material present at a star's birth determines its ultimate destiny—white dwarf, neutron star, or black hole.

Stars are born in vast, swirling clouds of interstellar gas and dust. Particles of dust and atoms of gas fall toward centers of condensation. Growing under an increasing infall of material, each condensation nucleus further extends its gravitational influence. Energy is radiated away at first, but later it is trapped in the interior as the mass increases. Heat and pressure build up in the accreting protostar, and after millions of years it reaches convective equilibrium. Shining dimly in the dust, the embryonic star releases more gravitational energy as it shrinks in size. Eventually convection gives way to radiation, and when the internal temperature reaches 20 million degrees, atomic furnaces turn on, and a star blazes forth.

## Red Giant

Stars "burn" by converting hydrogen into helium in thermonuclear reactions. A star of solar mass spends 99% of its lifetime burning hydrogen. But when the hydrogen is depleted, the core contracts and the star heats up, swelling to several hundred times its original diameter. The outer envelope of hydrogen is so far from the source of heat that it is relatively cool; it then glows as a *red giant.* Spread over a huge region of space, the star's density (except at its center) is so low that the air we breathe is three thousand times denser. Still, that is much denser than the surrounding vacuum of space. Aldebaran, Betelgeuse, and Antares glow brilliantly as red giants.

## White Dwarf

A helium core remains after the red giant's fuel is exhausted. Without nuclear burning, the helium core collapses, sending the temperature soaring to 100 million degrees. Other nuclear processes become possible at such temperatures. The helium itself may go through cycles of burning, producing carbon, oxygen, and neon. If the star is massive enough, the process continues, and heavier elements are formed. Finally, iron. And that's the end. The star is out of fuel.

Now 1/100th its original diameter, the star is the size of the Earth and a million times denser than water. Hot—white hot—the *white dwarf* is not easy to see, for brightness depends on surface area and the star is just too tiny to radiate much light. The companion of Sirius was the first white dwarf to be discovered. Procyon in Canis Minor also has a white dwarf companion.

A star resists further compression at the white dwarf stage because of its "degenerate" electrons. Ordinary atoms are mostly empty space. Negative charges, *electrons,* are held at great distances from the positive nuclei by electrical forces. However, under conditions of high temperature and pressure, electrons are free to move in closer to the nuclei. Each atom occupies less space, and the density of the gas increases a million-fold. Degenerate electron pressure halts further collapse, and the white hot ashes of the star take billions of years to cool.

## Neutron Star

If a star is between 1.4 and 2 or 3 times the mass of the Sun as it goes toward its final stages of evolution, then it may become a neutron star. Electron pressure cannot stop its collapse at the white dwarf stage; it overrides that limit and squeezes down to a thousandth of the diameter of the Earth. Electrons are driven into the nuclei of its atoms, neutralizing the charge and making them *neutrons.* Neutral nuclei can approach each other more closely in this situation than in the strongly electric atomic model, and pack tightly together into a neutron a million million times denser than water. Now 20 km in diameter, the neutron star is so dense that a tiny chunk of it placed on the Earth's surface would sink into it like a stone in water.

To visualize the process of a collapsing atom, think of the nucleus as a BB at the center of the 50-yard line on a football field. Electrons are on the top-row seats of the stadium, held at that distance by electrical forces. Most of the mass of the atom is in the BB, and most of the atom itself is empty space. Extreme temperature and pressure conditions force electrons into the central BB, and space collapses to a point.

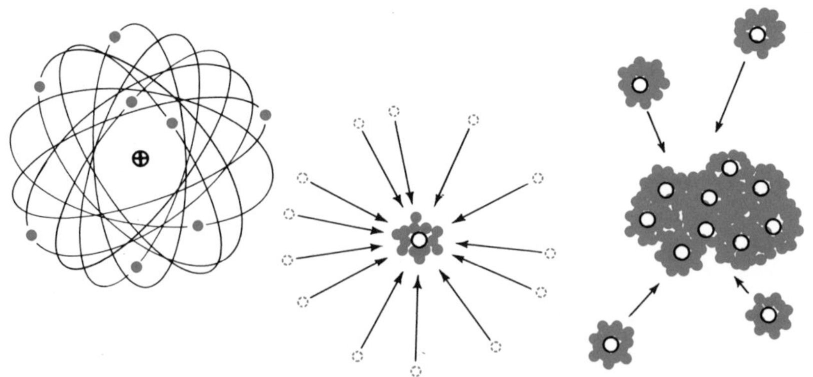

**Left: Diagram of a stable atom. A positively-charged nucleus is surrounded by a "cloud" of negatively-charged electrons, held in their orbits by electrical forces.**

**Center: Collapse of the atom. Under extreme conditions, the electrical force cannot hold electrons in orbit, and they drive into the nucleus, neutralizing its charge.**

**Right: Nuclei, carrying most of the mass of the atom, jam tightly together to form a neutron star of tremendous density.**

The Crab Nebula (M1) in Taurus. Near the center of this expanding gas cloud is a neutron star.

Neutron stars were predicted 30 years before any were discovered. The first to be identified optically is at the center of the Crab Nebula (M1), between the horns of Taurus the Bull. The Nebula is the remains of the supernova explosion; the Chinese and Japanese saw this supernova in A.D. 1054, reporting it to be bright enough to be seen in the daytime. The neutron star at the center of the Crab is a *pulsar,* a small object of tremendous density in rapid rotation, sending out bursts of radiation 30 times a second. Once predicted, now the neutron star has been "listened to" with radio telescopes and photographed with optical telescopes. Several are now known.

## Black hole

But if a star does not lose enough mass during its evolution and is still about 3 or more times the mass of the Sun, then neither electron pressure nor neutron pressure can stop its collapse, and it squeezes down into a "black hole" a few kilometers in diameter. So dense is it then—ten thousand million million times denser than water—and so strong is its gravitational force that nothing, not even light, can escape it.

**FORMATION OF A BLACK HOLE**

A massive collapsing star spins ever more rapidly, growing smaller and dimmer. When it shrinks below its critical radius, nothing can escape; space-time folds inward, and the star disappears from the universe.

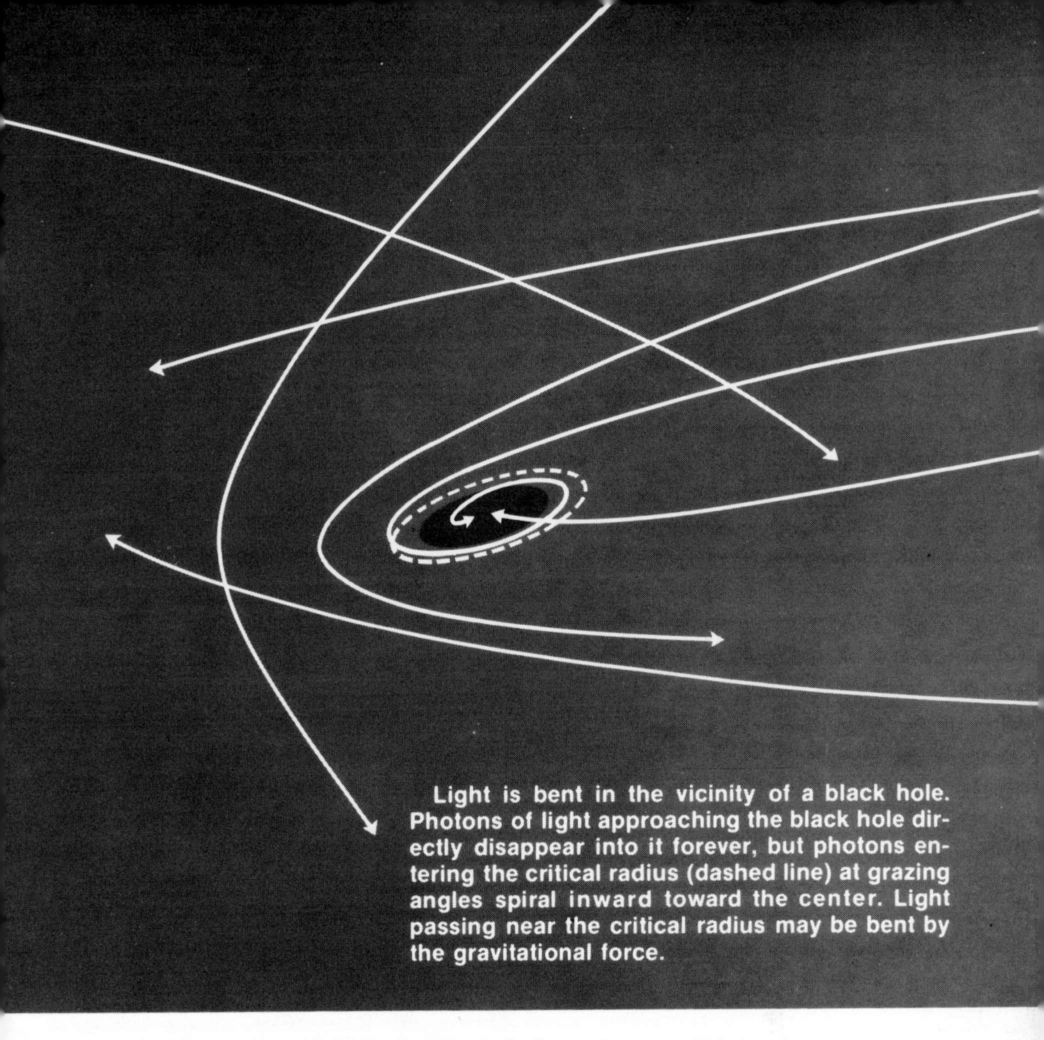

Light is bent in the vicinity of a black hole. Photons of light approaching the black hole directly disappear into it forever, but photons entering the critical radius (dashed line) at grazing angles spiral inward toward the center. Light passing near the critical radius may be bent by the gravitational force.

Space is strongly curved near a black hole. A beam of light passing at a grazing angle is bent toward that gravitational sink. Light entering within the critical gravitational radius disappears forever. Light trapped within the black hole cannot leave the "event horizon," for escape velocity is the speed of light.

A massive star could theoretically collapse into nothing and disappear forever from the visible universe, as if it had dug its own hole, jumped in, and pulled the hole in after it. But it would have to be perfectly spherical to become such a "singularity." CYG X-1, a strong x-ray source in Cygnus, may possibly be a black hole.

Some astronomers argue that if a star is squeezed out of existence at one place in the universe, surely it will appear somewhere else, possibly as a "white hole" through the Einstein-Rosen bridge. The idea of the white hole is pure speculation—a mathematical construct without support in physical theory.

OUR GALAXY

From a position in one of the spiral arms, we see our Galaxy as the Milky Way. The galactic center lies in the direction of Sagittarius and Scorpius. We look toward the Galaxy's edge to see Orion and the Gemini Twins.

## STARRY AGGREGATES: THE GALAXY

Stars that form the familiar groupings are close to us—100 to 200 light-years for those in the Big Dipper; 500 to 1,000 and more for those in Orion. Twenty-five to 50 times farther than Orion's stars, but in the opposite direction, is the galactic center. A hundred times the distance to that center is the Andromeda Galaxy; and a thousand times Andromeda's distance are the limits of the observable universe.

Stars are stages in the evolution of matter. Galaxies are starry aggregates of red giants, white dwarfs, black holes, pulsars, white holes, supernovae, and just ordinary stars—billions of them, bound together by gravity into a huge colony of stars.

We live near the outer edge of a spiral galaxy, some 30,000 light-years from its center. Rotation has flattened the Galaxy into a lens-shaped disk a hundred thousand light-years across, and ten thousand light-years thick. Star clusters swarm around the core of the Galaxy.

Mapping our Galaxy from within has been an extremely difficult task for astronomers—impossible, in fact, before the use of the radio telescope—because of the interstellar gas and dust that hides many of its stars. The dustiest region lies along the plane of the Galaxy. We see it as the "Rift" in the Milky Way running from Cygnus to

Sagittarius and beyond. Radio telescopes record the activity in that optically opaque region, mapping what we cannot see, far beyond the center of the Galaxy.

The galactic center lies in the direction of Scorpius and Sagittarius. Radiation from the center takes 300 centuries to reach Earth; that which is reaching us now had completed most of its journey—four-fifths of it—at the time civilizations were emerging on Earth.

We can get an idea of the immensity of our Galaxy by imagining a model: Let an orange represent the Sun. The nearest star, then, is another orange 3,000 km distant. The whole Galaxy, according to this citrus model, is an array of 100,000,000,000 oranges, separated by average distances of some 3,000 km, scattered over a lens-shaped region 60,000,000 km in diameter, and 6,000,000 km thick.

*A hundred billion stars in our Galaxy? It boggles the mind. We cannot imagine that number, but we can sense its magnitude by imagining the task of counting all the stars. At the rate of the human heart beat, about one per second, the counting would take more than a single lifetime . . . or 10 . . . or 30. It would take 50 lifetimes.*

*And then there's the Andromeda Galaxy with its twice as many stars—a hundred lifetimes of counting!*

Stars within a galaxy are born, reach maturity, then die. But the galaxy itself retains its shape, being balanced by the force of gravity that pulls everything inward, and by the force of rotation that opens it up.

Our Galaxy is one of twenty in the Local Group. Beyond that are other galaxies, clustering together in groups just as the stars within them do. Deep in space lie millions, perhaps billions, of other galaxies, each with billions of stars. And they all appear to be rushing away from each other, as if we're watching the remains of a gigantic explosion.

# UPON THE KNOWLEDGE OF CENTURIES

*From chaos*
*and the mists of nothing*
*came forth sky and fire,*
*earth and water,*
*and monsters—*
*then gods and men.*
*So sang the ancients.*

*We, piling measurement*
*upon the gathered knowledge*
*of centuries,*
*sense a different universe.*

*No longer singing*
*of gods*
*and heroic exploits of men,*
*we ask,*
*What is this that we're seeing?*
*One grand explosion?*
*Eternal perpetuation?*
*Pulsating cosmos?*

Ancient cosmologies deal with creative beings; modern cosmologies with creative forces. The personal universe expressed in the "I-Thou" relationship, and the predictable one expressed in mathematical terms—each has its truth value.

In the beginning, says Babylonian myth, the only living beings were Apsa (the Primeval) and Tiamat (Chaos). Nothing grew, for there was no ground. Nothing was named, for there were no namers. The waters mingled.

The gods, born of Chaos, were faced with the problems of separating Earth and Sky, giving names, and of ordering destinies. The attempt to bring order angered Tiamat. Filled with hate toward the gods whom she had borne, she roused up the Ancient Monsters to destroy them. One by one the gods went forth to appease her. Each failed from fear. Finally the lordly Marduk went forth, and the earth shook. Pulled in his chariot by four ferocious horses of high courage and swift pace, he rushed toward Tiamat, pierced her heart with his spear, and captured the Ancient Monsters.

Then Marduk split Tiamat's body, setting half of her in the Heavens above, and the other half below as the Earth. And he divided everything among the Heavens, the Earth, and the Abyss. He fixed the stars in the Heavens, caused the Sun and Moon to shine, ordained the year, and made man to live upon the Earth.

In the beginning, says a Greek story, was the black-winged bird, Nyx, hovering in the vast darkness. After eons of time, the bird laid an egg. When it hatched, Eros, the God of Love, flew forth. From the upper shell of the egg was formed Ouranos, the Heavens; from the lower shell, Gaia, the Earth.

Sky and Earth married. Their children were the Titans: Okeanos, Hyperion, Rhea, Tethys, and Kronos the cruel one. Kronos hated his father, Ouranos, and wounded him so severely that Ouranos and Gaia were forever separated from each other. Kronos married Rhea, and their children were different from the Titans. Zeus, their most famous son, went to live on Mt. Olympus.

Now the Greeks lived in a land which they believed to be the center of the universe. Flowing around the Earth was the River Ocean. In spite of storm or tempest, its flow was always tranquil. From it, all the rivers of the world received their water.

Beyond the cavernous mountains to the north from which chilling winds blew, lived the happy Hyperboreans, dwelling in eternal spring, exempt from disease, old age, and the toils of life.

The happy and virtuous Aethiopians lived on the south side of Earth. So favored were they, that the Olympian gods often descended from their lofty heights to join them in their feasts.

The blissful Elysian Plain was at the western margin of Earth. Mortals favored by the gods were transported here without experiencing the sting of death.

Out of the ocean on the eastern side of the world rose the Dawn, Sun, and Moon, giving light to gods and men. The stars rose and set, too, except those near the Great Bear that continually circled the heavens.

In the beginning, says Icelandic myth, was the Place of Fire and the Place of Fog. Between them was the Yawning Gap. Out of the Mists of Yawning Gap came the first two beings—Ymir the Giant, and Audhulma the Cow. From the feet of Ymir sprang a race of Giants. Audhulma the Cow licked an ice cliff. Gradually a figure more shapely than Giant's began forming. When the figure—Buri, a man—began to breathe and move, he separated himself from the ice cliff as a living being. He married a Giantess, and they lived in peace with Ymir and his children.

But multiplying brought strife, and the Giant Ymir was slain. His body, flung into Yawning Gap, filled the chasm. And of his bones are mountains made; and of his teeth, the rocks; and of his hair, the grasses of the meadows and the trees of the forests; and from his hollow skull, the sky. From sparks and clouds of flame came the Sun and Moon and all the stars of night.

In the beginning, says the Haggada, the Most High created the heavens and the Earth, the light and the darkness. Then He took a stone and threw it on chaos, and it became the Earth. On the second day, He created the angels. On the third, the plants. But along with the plants He also created iron to fell the Cedars of Lebanon lest they exalt themselves unduly. On the fourth day He created the Sun, Moon, and stars. And on the fifth day, fishes and birds. On the sixth day, He created animals and also Man.

In the beginning, says a Chinese story, there was the Yang-Yin—the light-dark, the heat-cold, the dry-moist. The subtle went upward, forming the Heavens; the gross went down, forming the Earth. From the interaction of Yang and Yin, the active and passive, the male and female, came the seasons and the flowering of the Earth. Warm effluences from Yang produced fire that formed the Sun. Warm effluences from Yin produced water that formed the Moon. The Sun, interacting with the Moon, produced stars. The combining of Yang and Yin produced all creatures, all things. The power of Yang and the receptivity of Yin is All. The Heavens were adorned with Sun, Moon, and stars; and the Earth received dust, rain, and rivers.

But it is also said that long before Yang and Yin were separated, P'an Ku created the universe, chiseling planets and stars out of a cliff representing chaos. Dragon, Unicorn, Tortoise, and Phoenix accompanied him as he worked. When he died, his limbs became the four corners of the Earth; his head, the mountains; his flesh, the soil; and his breath, the wind. The task of creation took 18,000 years.

In the beginning, says Egyptian myth, Shu, the God of the Air and Sunlight, separated the Sky Goddess Nut from Seb, the Earth. The body of Nut arches across the heavens, her feet touching one horizon, her fingertips the other. Each day the Sun God Ra, seated in a boat, rides across the sky. The Earth at the center of the universe is a flat platter with a corrugated rim, floating on the waters of the abysmal regions of the underwold. Heavenly bodies are the lights on the celestial vault—a sight so awesome that it causes a "welling up" in the souls of human beings.

In the beginning, says a Mexican myth, all was silence. All was motionless. All breathless. Everything was separate. Nothing joined. The boundlessness of the sky matched the quietude of unruffled waters. All expectant. Then the waters drew back and the land appeared. Its surface hardened and it became sown with seed. Forests and mountains appeared, and there was flatness no more. The gods made creatures and placed them on the Earth. And they also made human beings, endowing them with intelligence.

In the beginning, says Indian myth, Surya the Sun God drove his chariot across the sky, pulled by seven magnificent horses. Surya married the wise goddess Sanjna, and their children were the first man and woman on Earth. Blinded by her husband's brilliance, Sanjna left him. She wandered for a long time by herself. Surya searched, and eventually they were reunited. Then he gave some of his brilliance to the Moon and to the stars and to other gods so that Sanjna would not have to flee again. By giving away some of his power, he gained strength and stability.

In the beginning, says a Japanese story, was the great swirling ocean mass. From it appeared a pair of gods who, upon descending to Earth on a rainbow bridge, became the parents of the world. From the left eye of the Great Father was Amaterasu the Sun Goddess born. From his right eye came the Moon and the stars. And from his nostrils came forth the powerful Storm God, Susanowo.

Susanowo, the irrational, ravaged the land, destroying trees and crops, breaking down that which was built up by Amaterasu. Seeking to end the continual strife between them, they exchanged tokens of their trust in each other. For a while, all was well. Susanowo stayed in his realm, the sea; Amaterasu had her way upon the Earth and full crops were harvested.

But after a time Susanowo left his domain. Again he stormed across the land, throwing down temples and scattering the people. No more would Amaterasu look upon that destruction. Shamed, she went to hide in a cave, and all was darkness on Earth. Deprived of the light in the sky, the gods consulted together on how her beneficence might be restored. They gathered outside the cave in which she hid, and lighted fires and sang. The Goddess of Laughter, Uzume, danced on an upturned tub. Her dancing and the laughter made all the other gods happy, and they, too, laughed. Amaterasu, alone in the cave, heard the sound of merriment. Curious, she ventured out, and light immediately filled the world. The gods were overjoyed and welcomed her back, glad to have the resplendent Sky Goddess lighting the Earth once again.

In the beginning, says American Indian myth, the Creator made the world, starting in the east. Coming westward, he gave different languages to the people in each region he created.

Tribes could not talk with each other because of their various languages. But they had one thing in common: they were not at all pleased with the Creator's creation. The sky was too low, and tall people bumped their heads against it. Those who climbed too high in trees went right into the sky world.

So they met in council. A wise man said that they could push the sky up higher if they'd all work together. From tall fir trees they made sky-pushing poles. The great day came. Wise men shouted, "Ya-hoh!" and everyone pushed. The sky moved upward just a little. Time and time again they shouted and pushed in rhythm until they raised it to where it is today—too high for head-bumping, and too far above the trees for easy access into the sky world.

In the beginning, according to African myth, were the twin gods Lisa and Mawu, brother and sister. Brilliant and fierce was the brother, Lisa, the Sun, who lived in the east. Gentle and beautiful was the sister, Mawu, the Moon, who lived in the west. Regularly they drew close to each other in the heavens, but they had no children. One time Mawu dimmed Lisa's brilliance during an eclipse, and they had seven sets of twins, all gods and goddesses. The seven pairs became the planets and stars. Now, whenever there is an eclipse, people on earth know that Mawu and Lisa are making love.

In the beginning, says a Polynesian story, the Sky God, Wakea, and the Earth Goddess, Papa, held each other in tight embrace. Caught in between, the children of the gods had but little room in which to move. All was darkness. There was no day; only night. Nothing could grow, nothing could ripen, and there was no maturity.

The children of the gods had heard of light, but they had experienced only darkness. They wanted to move, to explore freely, to know the light, and to attain their own identities. Kane, the greatest of the young gods, suggested that they force apart mother and father. One by one they tried, and one by one they failed. At last, Kane lay on his back, pressed his shoulders against his Earth-mother and his feet against his Sky-father. He pushed and pushed, exerting the utmost effort. Slowly Earth and Sky began separating. Light flooded between them. "What have we done that you treat us so?" cried Papa and Wakea. Tears that the Sky-father sheds as he bemoans the separation are the gentle rains that fall from trade-wind clouds. And the tears that the Earth-mother sheds as she bemoans her fate are the gentle mists that roll up in mountain valleys. And in the new light of the Earth, the children of the gods explored widely over the sea and up into the sky.

In the beginning, says an Australian aboriginal myth, the birds were arguing over the excellence of their chicks. They fluttered and chattered and scolded each other. Finally in anger one bird took the eggs from another's nest and threw them into the sky. There, on a pile of sticks gathered by the sky people, they shattered and splattered, bursting into a brilliance that revealed for the first time the beauty of the Earth beneath.

Ancient man stated his cosmology in myth. We state ours in scientific terms. Cosmological models, then, have shifted over the past few thousand years from those based on mythological outlook, to those based on rational and relativistic thought.

## THE PERSONAL UNIVERSE

Less than ten thousand years ago, the last glacier retreated from northern Europe. Agriculture could then begin, and man derive from crops the energy he needed to sustain life.

Successful farming is tied to knowing when to plant. A clue is the Sun's position along the horizon, which varies with the seasons; another is the annual appearance of certain stars. Agrarian man noted these clues and contrived calendars to record that annual rhythm.

Superimposed upon seasonal cycles, though, were unexplainable times of dry and wet, of hot and cold, of flood and drought—unexplainable except as the whim of the gods or as man's punishment for wrongdoing. The rage of the gods manifested itself in thunder and lightning, in angry waves stirring up the sea, and in violent storms venting fury upon the land. The universe was personal, full of power. Above all, irrational. Myth explained the inexplicable with a logic consistent with its own precepts.

## THE RATIONAL UNIVERSE

A wave of intellectual brilliance broke in the 6th century B.C., bringing new ways of looking at the world. Buddha appeared in India; Lao-Tse and Confucius in China. The Gilgamesh Epic was written on clay tablets in Babylon, and the Jews were rebuilding the Temple. Citizenship was conferred upon all the free inhabitants of Athens.

The Ionian philosophers in Greece began applying reason to the understanding of natural events. The Earth, to be sure, was still an oyster shell floating on the sea. But the model was refined through an intellectual process of observing, questioning, applying mathematics, and arguing. Pythagoras may have originated rational cosmology, but the quintessence of cosmological thought was achieved five centuries later in Ptolemy's model of the universe.

The wave receded. For a long time all was quiet. Gathering

momentum, it broke once more, this time at the end of the 15th century. New theories about the size and shape of the Earth were tested by voyaging into the unknown. Columbus found no edge to the Earth, nor did he expect to. Still, it took another thirty years for sailors to find a way around the extensive American landmass, proving the world to be boundless and finite. Proof of the roundness of the Earth, however, in no way changed the prevailing view that the world was the center of the universe.

Half a century after Columbus, Copernicus cautiously put forth his sun-centered model of the universe. Fifty years later, Kepler, using Tycho Brahe's precise data on planetary motion, worked out the mathematics that supported the Copernican model. Much to Kepler's surprise, he found that planets move in ellipses, not in perfect circles.

A great step toward new cosmology was the invention of the telescope. Galileo, in 1609, used the telescope as a scientific instrument. With it, he found that the Sun had spots, and that the Moon was rough and mountainous like the Earth. He also saw four moons circling Jupiter—proof that the Earth was not the only center of rotation in the universe.

Kepler had found a relationship between a planet's orbital speed and its distance from the Sun, but he had no theory to go with it. Galileo had delved into the mysteries of gravity by rolling objects down inclined planes, but he missed the big picture. It was Newton who put it all together, providing mathematical validation of the empirical discoveries of his predecessors in his theory of gravitation, 1687. He gave generous credit to them: *If I have seen further, it is by standing upon ye sholders of Giants.*

A triumphant validation of the universality of Newton's gravitational theory took place a century later with the prediction of the planet Neptune, then, finally, its discovery in 1846.

## BREAKING THROUGH THE CELESTIAL SPHERE

Are the "fixed stars" really fixed? Immovable? Are they close? Infinitely far? Edmund Halley, Newton's friend and helper, discovered in 1718 that Sirius, Aldebaran and Arcturus were no longer exactly where Hipparchus had placed them on his charts two thousand years before. Halley presumed that they had moved, and he set about to detect motions in other stars.

Halley is perhaps most noted for his prediction of the return of the comet that now bears his name. An even more important contribution to science was his suggestion that the transits of Venus in 1761 and 1769 would be excellent times for determining the Sun's distance to a high degree of accuracy, for estimates varied by about 30% at the time. By making observations of Venus' transit from widely scattered places on Earth, the uncertainty could be cut to 0.2% of the true value. Scientific expeditions set out to observe the phenomenon. Captain Cook's voyage to Tahiti in 1769 was a consequence of that coordinated scientific investigation.

Late in the 18th century, William Herschel was convinced, as were most astronomers, that stars were not stuck to a celestial sphere but were scattered throughout space at various distances from the Earth. He built a huge telescope 40 feet long with a 4-foot mirror. With it, he estimated the relative distances of stars according to brightness. The dimmer the star, he reasoned, the farther away it is; one that is half as bright as another must be four times the distance. He found stars occupying a space shaped like a "grindstone," and he discovered "island universes" as well.

Variations in light output from Mira and Algol indicated stellar individuality to Herschel. Stars traveling around each other—binary stars—were further evidence of the depths of stars in space. He brought descriptive astronomy to its peak, but that was not his goal; in 1811 he said, "A knowledge of the construction of the heavens has always been the ultimate object of my observations." That possibility had to await the development of new instruments.

Bessel measured the distance to 61 Cygni in 1838. Using the sun-earth distance as a base line, he measured the shift of 61 Cygni, finding its distance to be a half-million astronomical units. Modern measurement puts it at 11.1 light-years—about 700,000 AU.

## STELLAR CHEMISTRY

The 19th century brought not only knowledge of the distances to stars, but also an understanding of the message contained in starlight. Fraunhofer found, in 1814, that the rainbow spectrum of the Sun contained some 600 dark lines. He had no explanation for what he observed. Thirty years later, Kirchhoff showed that the dark lines in the spectrum corresponded to bright lines that appear when certain elements are burned in the laboratory, indicating the presence of these elements in the Sun's atmosphere.

The elements that human beings are made of are the same as those in the Sun, the same elements found in the most distant stars. Every element has its characteristic spectral lines, and chemists have found an orderly system of elements in the universe, increasing in complexity from hydrogen to uranium.

Each star broadcasts what it is made of. The spectroscope, using a prism or diffraction grating, spreads starlight out into its rainbow colors. The spectrograph photographs the image and records the spectral lines, indicating the chemical composition of the star.

Stellar temperature, velocity, luminosity, magnetism, and spin rate can be inferred from the spectra.

## REVOLUTION

Classical physics reached its peak during the 19th century. Kinetic theory and the conservation of energy had been so often validated that science achieved a dogmatic position. Its structure, so perfect; its ideas, so beautiful; its logic, so impeccable; and its success so brilliant that it seemed as if all the laws of the universe had been discovered. Only a little polishing was yet needed.

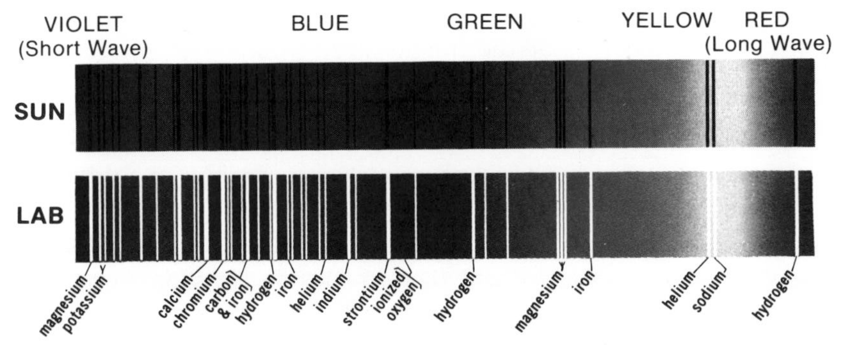


**When it was learned that the spectral lines of the Sun and stars correspond with those of elements burned in laboratories, it became clear that celestial objects contain the same elements that are present on Earth.**

All was clear, except, as Lord Kelvin put it, "two small dark clouds on the horizon" which he thought would soon disappear. One cloud was the failure of experiments to find an "ether wind" through which light waves were propagated. The other, the "ultraviolet catastrophe," in which the strongest radiation from glowing objects was not in the ultraviolet, as had been predicted by classical theory.

But Kelvin's tiny clouds did not float away. Instead, they grew in size. Within them were the seeds of two revolutions—quantum and relativity—that shook the foundations of classical physics.

By the end of the 19th century, no one yet knew the size of the universe and how it worked, what kept the Sun burning, or what was going on in stars, dust clouds, and galaxies.

Max Planck found, at the turn of the century, that radiation from glowing bodies could best be explained by assuming that light travels in bundles—quanta—not waves. According to classical physics, though, light is a wave phenomenon. Experiments that continually confirmed the wave theory culminated in Maxwell's wonderful equations. Even Planck had trouble in accepting what he had discovered because it was so against prevailing belief. Einstein, though, built quantum ideas into his photoelectric theory in 1905.

Rutherford developed a model of the atom in 1911 in which negative particles surrounded a positively charged nucleus. Applying quantum theory to Rutherford's model, Niels Bohr developed one in which electrons emitted photons of light when jumping from one orbit to another. Limitations in Bohr's model led de Broglie in 1922 to his theory that the quanta are accompanied by guiding waves.

Is light a particle or a wave phenomenon? Both. It's an unpicturable entity that seems to be a particle when we make experiments of one type, a wave when we make experiments of a different sort.

The wave-particle duality brought uncertainty into science: the very act of observing disturbs that which one is trying to see. In order to "see" an electron, we bombard it with a photon of light. But that gives the electron a jolt that sends it in some unpredictable direction. Use a longer wavelength of light with lower energy to give the electron less jolt, then its clearness fades. According to Heisenberg's *uncertainty principle* we cannot, at the same time, know the particle's precise position and its motion, so we cannot predict where it will be later on. Classical physics was based on predictability and causality; this uncertainty simply would not fit.

Kelvin

The age of classical physics ended. Not abruptly. For there were giants in those days—Einstein, Schrödinger, and de Broglie—who supported the determinism of classical notions. The Earth shook as they argued with advocates of the quantum theory—Born, Dirac, Heisenberg, and Pauli—who supported the probabilistic nature of science. Because quantum theory does not have the purity of classical physics, some doubted it. Einstein contributed heavily to quantum theory, but he could not accept its base in probability as anything but incomplete. "God does not play dice," is a famous remark he made many times. He continued to look for deterministic law that could predict precisely what would happen in a system whose parameters are known, and where certainty, not probability, is fundamental.

But, since probabilities flow from billions of uncertainties, the quantum theory works, even though it is not satisfactory to some. The theory has led to a new knowledge of the atom. The neutron was discovered in 1932. Fermi saw that the neutron could be used to split the uranium atom, and the terms *fission* and *fusion* entered the common vocabulary. Quantum theory has led to the invention of the radio telescope and television. It has made orbital flight possible, as well as trips to the Moon. And it has led to an understanding of what powers the Sun and other stars.

Hydrogen is converted into helium in the "burning" of a star. Only a small amount of mass (0.7%) is used up, but an astounding amount of energy is released in the process. Einstein's celebrated equation predicts the value:

$$E = mc^2$$

Expressing the mass in kilograms and the velocity of light in meters per second, the energy produced is related to the mass by a factor of 90,000 trillion. So great is the energy released, that a mass of two grams (about that of a dime) converted into energy produces more than 6,000 kilowatts—enough to power the average household for an entire year.

Einstein

Each second, more than 4,000 tons of matter are converted into energy by the Sun. But the Sun is so big that even at that rate it can continue to expend energy for billions of years before burning out. When all the hydrogen is used up, 99.3% of the original mass will still remain.

## THE RELATIVISTIC UNIVERSE

Would Newtonian physics hold in the vast distances of space where massive objects are moving at three-quarters the speed of light? Or in laboratories when particles are traveling in magnetic fields at 99.9999% the speed of light? Here, Newtonian physics gives way to relativity theory.

The size of the universe was not clearly understood at the time relativity was proposed: astronomers generally were of the opinion that the Milky Way comprised the whole universe. Some, though, speculated that the dark patch in the Southern Cross was a window through the Milky Way to the universe beyond.

What, really, are we looking at? The Shapley-Curtis debate of 1920 dealt with that question. The two notable astronomers debated the Sun's position in the Galaxy, and whether or not there was just one galaxy or many. Each was right on a point; each wrong on another. Shapley's observation of the bunching of globular clusters to one side of the Milky Way was evidence for him that the Sun was not at the center of the Galaxy. He was right on that point, but he argued against the existence of external galaxies. Curtis, on the other hand, was wrong on the position of the Sun in the Galaxy, but right, as proven later, on the idea of many "island universes."

Einstein was interested in finding the universal laws of nature, for he was aware of the anomalies that indicated something was wrong with the approach of classical physics. Assuming the speed of light to be constant, he reformulated the laws of physics for systems that are at constant speed relative to each other, and published the *special theory of relativity* in 1905. Then he developed the *general theory of relativity* (1916) that would apply to systems in acceleration, and found that it is not necessary to think of gravitation as a force.

Time, in Newtonian physics, is absolute. It is measured by rhythmic events: the swinging pendulum, the rotating Earth, the vibrating cesium atom. But in relativity, neither time nor motion is absolute. What moves—the Earth, or the space ship going to the Moon? From the astronaut's point of view, it's the Earth that is moving. All motion is relative, the observed motion depending upon one's observation point. But time varies with speed. As speed increases, time slows down. For the astronaut, Earth clocks are running slowly; but from the terrestrial observer's point of view, it is the astronaut's clock that is running slowly, despite the finest adjustment.

Einstein expressed gravitation as a manifestation of the geometry of space-time. The geometry is affected in the presence of massive objects. A ray of light does not go in a straight line. It follows the form of space, the *null geodesic*. The curved-space hypothesis was tested during the eclipse of 1919. A star, predicted to appear even though it would be behind the Moon-covered Sun, *did* appear. Curvature was slight—only a fraction of a degree—but it supported the hypothesis. The concept of the warping of space-time was further confirmed when radio signals from Mariners 6 and 7 were found to take 100 microseconds longer to reach Earth when on the far side of the Sun than they would have if the Sun had not been there.

Einstein went on to develop a model of the universe. Finite in-

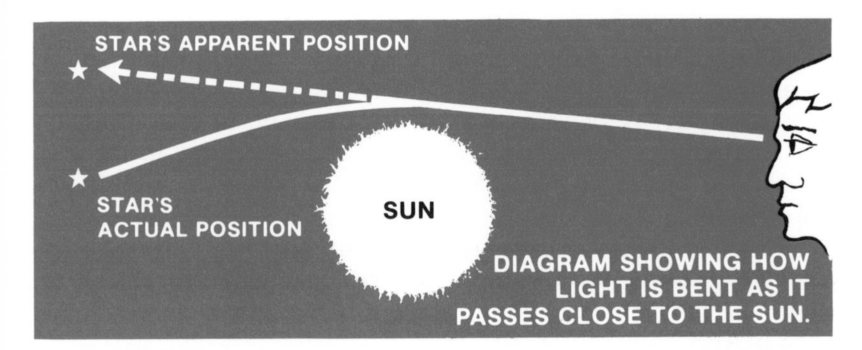


size yet unbounded, the space-time curvature of his relativistic model was like the surface of a sphere: travel far enough in one direction and you come back to where you started. De Sitter came up with a negative-space model in 1917 in which the universe was not only unbounded but infinite in size. Space-time had an intrinsic structure of its own, independent of the presence of matter.

Friedmann pointed out the instability of both models. He worked out a mathematical model in 1922 whose radius is constantly changing with time—an expanding universe. According to this model, light leaving a distant galaxy would have its spectral lines shifted toward the red if the galaxy was moving away, and toward the blue if moving toward us. Confirmation of the expanding universe came in 1930 with Hubble's discovery of the *redshift* phenomenon: the farther a galaxy, the faster it is receding, and the greater its redshift.

## BIG BANG

In the beginning, according to the "big bang" theory, all the material of the universe was gathered together in one place. No one knows where it came from or why it was there. The initial conditions can only be described, not explained.

Suddenly it exploded. Parts of it went to form spiral galaxies, globular clusters, dust clouds, and pulsars. Temperature dropped rapidly from the initial hot 500 billion degrees to a mere billion in a matter of five minutes. The intense heat of the big bang, in which helium formed, may have left a permanent mark in the universe—a field of radiation that has almost certainly been detected by the radio telescope. Launched in a blaze of light some 12 to 15 billion years ago, the universe continually evolves from its radiation-dominated beginning.

## STEADY STATE

In the beginning, says the steady state theory, there was no beginning. There always was; there always will be. The universe is infinitely old, infinitely large, not evolving. It looks today as it always has. Its average density remains constant, in a steady state. Out of nothing, hydrogen is continually being created in the vast interstellar spaces at just the right rate to replace material that forms stars and galaxies, then it drifts out into the endless void.

How can matter be created out of nothing? That's no more difficult to explain, say the steady statists, than how, according to big bang theorists, everything in the universe got together in the first place.

## MID-CENTURY DEVELOPMENTS

The radio window on the universe was opened in 1945. Twelve years later (October 4, 1957) Sputnik went into orbit. Twelve years later (July 20, 1969) man set foot on the Moon.

Radiation from all over the universe is constantly streaming past the Earth. Most of it is screened out by the atmosphere, but light and radio waves, along with some infrared and ultraviolet, do get through to us. We feel the infrared as heat, our skin responds to the ultraviolet, and our eyes respond to light waves so short that 2,000 of them can fit across the period at the end of this sentence.

The optical window was the only one open to the universe until the middle of the 20th century. Methods of detecting radiation coming through the atmosphere at radio frequencies had not yet been developed, nor had technology arrived at ways of hefting telescopes into space far above the aberrations caused by the atmosphere. Modern instruments are gathering data in infrared, ultraviolet, radio, x-ray, and gamma radiation. Computers are processing the data; astronomers are interpreting the results and developing new theories. The vocabulary of the new cosmology includes such terms as pulsars, quasars, and black holes—terms unknown before the 1960s. Now that the electromagnetic spectrum is completely open, some astronomers think that we have actually detected the boundaries of the universe.

Radio waves are of much longer wavelength than light waves. Due to this long wavelength, large collectors are needed for resolution. The 250-foot parabolic dish at Jodrell Bank, England, is the largest completely steerable radio telescope in the world. A somewhat steerable one, four times that diameter, nestles in a natural

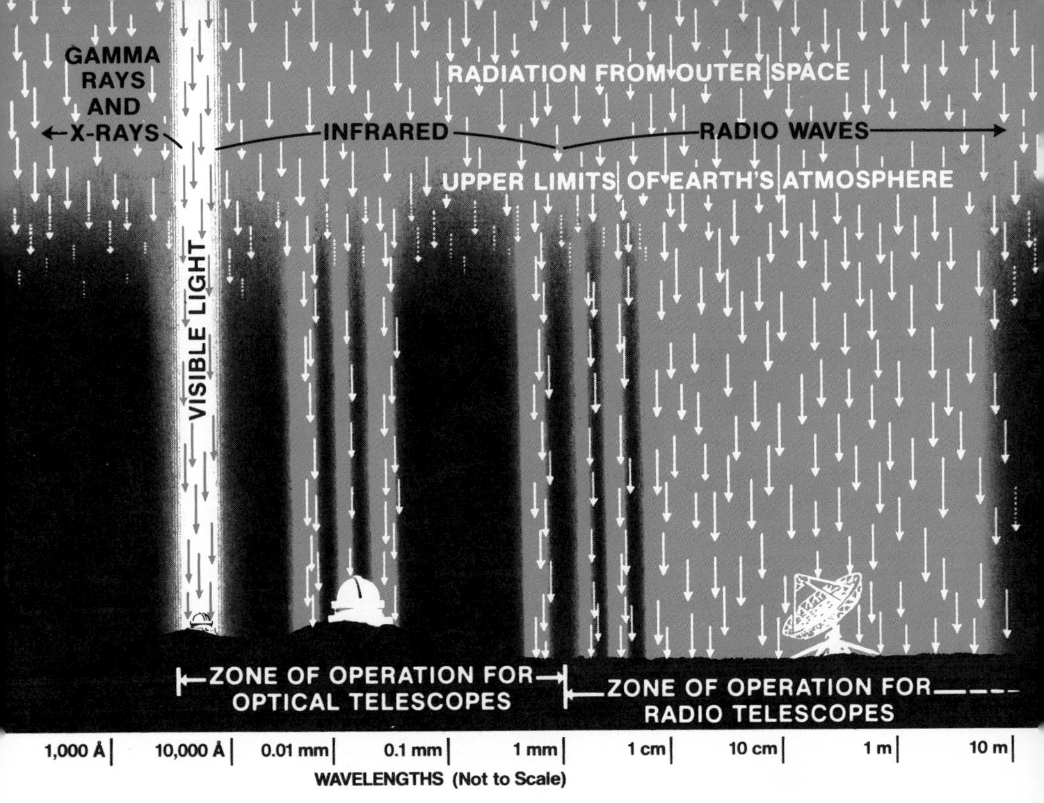


**WINDOWS IN THE EARTH'S ATMOSPHERE**

**Dark areas indicate where Earth's atmosphere blocks radiation from space; blue and white areas indicate windows in the atmosphere through which radiation reaches Earth, where ground-based instruments are able to gather data.**

bowl in a rugged limestone region of Puerto Rico. The Mills Cross radio telescope in Sydney, Australia, is an x-shaped array of detectors (each a bowl or dish) 1,500 feet long, set at right angles to each other.

Resolution—the ability of a telescope to separate sources—is a great problem when using the long waves of radio frequency. The problem is solved by setting instruments great distances apart, even on opposite sides of the Earth, and linking them electronically.

Why, after a thousand years, is the Crab Nebula still so visible? What keeps its "wisps of brightness" bright as they speed outward a thousand kilometers each second? The radio telescope found a heavy flux of x-ray at the center of the Crab. Polarization of light indicated strong magnetic fields. At first a neutron star seemed to be the answer—a small discrete source less than 20 km in diameter that should wink out suddenly if the Moon moved in front of it.

MOON OCCULTING PLASMA CLOUD OF CRAB NEBULA

But measurements during the 1964 lunar occultation of the Crab showed slow fading and slow restoration, suggesting that a large plasma cloud was being observed, instead of a pin-point star.

The radio telescope found, in 1967, a *pulsar* at the center of the Crab Nebula sending out bursts of energy 30 times a second. The pulsar turned out to be a neutron star after all! A rapidly rotating object a million million times denser than water, it produces strong magnetic fields, whirling charged particles into the glowing cloud, and keeping the Crab bright through *synchrotron radiation.*

The radio telescope entered molecular astronomy in 1968 with the discovery of interstellar water, ammonia, carbon dioxide, formaldehyde, methyl alcohol, cyanogen, and many other compounds. Such molecules form the basis of life on Earth. Carbon and oxygen—elements of which we are made—were forged in the hot interiors of stars no longer in existence.

But the radio telescope is of no help when it comes to neutrino astronomy. The neutrino, a particle of zero mass traveling at the speed of light, is the most elusive of all the atomic particles, and almost impossible to detect. Produced in thermonuclear reactions at the center of the Sun, the neutrino travels through the half-million miles of solar plasma as if nothing were there. So elusive is the neutrino that some of them could theoretically travel through solid lead from here to Castor, some 50 light-years distant, without striking a thing.

A strange neutrino-capturing "telescope" looks at the Sun from a mile beneath the Earth's surface in South Dakota. It's a 100,000-gallon tank of cleaning fluid buried deep within the Homestake Mine. The overlying burden of rock screens out cosmic particles, but neutrinos go right through. Even through the Earth. Occasionally, a neutrino strikes a chlorine atom in the cleaning fluid, producing radioactive argon. The amount of radioactivity in the tank is an indication of what's going on at the center of the Sun where neutrinos are being produced. If the neutrino output could be determined to within 50% accuracy, we could know the Sun's central temperature to within 10% of its true value.

Another "telescope" that never sees the sky searches out gravity waves that originate in the center of our Galaxy. It's a solid aluminum cylinder a meter in diameter, 1.5 meters thick, weighing 3.5 tons. Wrapped around the center of the cylinder are quartz crystals that are wired into a recording system. As gravitational waves distort the cylinder's shape ever so slightly, the crystals respond, producing tiny electrical currents that are fed into high gain amplifiers. The instrument is extremely sensitive—so sensitive that to minimize the effect of local disturbances, two stations have been set up: one in Maryland and another in Illinois. Only the "spikes" that appear on both recording charts are considered significant. Two or three spikes a week indicate that an amount between a hundred and a thousand solar masses are being converted into gravitational energy each year at the center of the Galaxy—a violent place, indeed.

## GALACTIC FORM AND SUBSTANCE

Stars in the core of our Galaxy are generally cooler and redder than the young blue stars in its spiral arms. The galactic center seemed placid until recently, when infrared measurements showed it to have a tiny hotspot 0.3 light-years across. Ultraviolet light (far more energetic than infrared) also indicates a hot core for our Galaxy, and for the Andromeda Galaxy as well.

Working together, the radio and optical telescopes, and other instruments, are finding a tremendous variety, variability, and violence within galaxies. Ranging in size from thousands to hundreds of thousands of light-years across, galaxies also come in various shapes: ellipticals, spirals, and irregulars. Irregulars contain the most interstellar gas, and new stars are forming in their dust clouds. Ellipticals contain hardly any gas; they're made of old stars and not much is going on. Ellipticals can be almost spherical in shape, while spirals are flat. All galaxies rotate, spirals rotating the most rapidly. Some galaxies are even exploding. And they're all moving away from each other at speeds proportional to their distances.

How can galaxies explode when the stars within them are so tremendously far apart? The average spacing of stars in our region of the Galaxy is four light-years—about 30,000,000 solar diameters. Or, according to our "citrus model," stars are oranges placed 3,000 km apart. How could anything *that* tenuous explode?

The radio telescope has located galaxies that are in states of extreme disorder. Huge jets of matter hurtling outward from the Virgo A radio source (M87) indicate a galaxy in distress. Faint streamers of ionized hydrogen shooting out thousands of light-years from the core of M82 in the Big Dipper, show it to be an exploding galaxy. Cygnus A, once thought to be two colliding galaxies, is now recognized as a galaxy tearing itself apart with a force equivalent to ten billion Crab supernovae. A galaxy in Corvus (NGC 4038-9) is doing the same. Some galaxies seem to be giving birth to others: unusual clouds connect a group of five in Serpens. And a galaxy with a bright center (3C120) lost half its brilliance in seven weeks.

One in a hundred galaxies has a center so bright that we see it as a star. Such "seyfert" galaxies may be the link between the quasar and the true galaxy. Quasars are intense radio emitters. Optically, they look like stars, producing as much light as a hundred galaxies; yet they're small, having diameters a hundred-thousand times smaller than galaxies. Rapid fluctuation in both light and radio wavelengths indicates small diameters. A 16-fold variation in the brightness of one quasar is attributed to strong magnetic fields. Another quasar doubles its brightness suddenly every twelve weeks, then quickly declines. Broad spectral emission lines suggest extremely hot gases and great turbulent velocities.

Perhaps quasars are chain reactions of supernovae, a million of them exploding within a hundred years of each other. But they might also be objects thrown out of galaxies at speeds in excess of 1,500 km/second. Or they may be produced by unknown mechanisms related to the energy released in gravitational contraction. Or in the encounter of matter with antimatter.

Swedish astronomers have proposed a theory of the universe based on antimatter. Every particle, according to quantum mechanics, has its antiparticle—a basic symmetry of nature that is confirmed in laboratory experiments. We happen to live in that part of the universe where the atom consists of a positive nucleus surrounded by electrons of negative charge. Perhaps worlds of antimatter exist where the reverse is true, with positive electrons surrounding negative nuclei. A star of antimatter would look just the same as a star made of matter. Observationally we could not tell the difference. If antimatter reaches the earth, it can exist only briefly, for in the encounter with matter both are completely annihilated.

In the beginning, says the Swedish theory, was a cloud of matter and antimatter a trillion light-years in diameter. Shrinking under gravitational force to a thousandth of its original diameter, its density increased a billion-fold. Collisions of matter with antimatter became frequent enough eventually to produce radiation pressure sufficient to overcome gravity. Galaxies moved through the center and out again. We're looking now at an expanding phase.

## CONTRASTING COSMOLOGIES

Cosmology is the search for the form of the universe. Modern cosmological models, unlike ancient ones, can be tested, for they are based on measurement of the motions in the universe and the distribution of matter. Imagination in company with detailed measurement is central to the spirit of modern science.

The universe evolves, or it does not evolve. Either it all started with a big bang and is continually changing, or it always was as it is and continuous creation keeps it so. The big bang and the steady state cosmologies are two contrasting points of view giving rise to hypotheses that can be tested.

Most cosmologies are based on the *cosmological principle*—that we are not in any unusual place in the universe, and that what we see is representative of the whole. A few theories that are not based on that principle involve hierarchies of clustering, or changing physical constants with time, such as increasing the masses of atoms.

If the universe is expanding (and how else can redshift be explained?), then it must have started at a point in time. All galaxies must be of the same relative age, having been created in the big bang. The farther into the distance we look, the farther back into time we see—back toward creation when galaxies were young. Steady state theory, though, predicts young and old galaxies side by side.

According to the big bang theory, the universe was denser billions of years ago than it is now; expansion is thinning it out. Galaxies seen at great distances (and thus far back in time) should be closer together than the ones near us. A great number of weak cosmic sources picked up by the radio telescope from remote regions of the universe are evidence supporting the evolutionary cosmologies' concept of greater ancient density. Steady state theory predicts that the average density never changes.

Quasars also challenge the steady state theory. Most quasars (if their redshifts are truly cosmological) are at a tremendous distance; they are among the most distant objects yet observed. So they must have come into existence long ago. Steady state theory, on the other hand, would have them all over the universe, some forming even now, and nearby.

But the greatest challenge to steady state is fireball radiation. Originally, the fireball of the big bang was hot and opaque, like the interior of a star. A million years later it became transparent as termperature fell to 3,000 degrees Kelvin, and radiation, then no longer absorbed, could pass freely into the universe. Redshifts would by now have transformed that temperature to about 3 degrees Kelvin—low, from a laboratory point of view, but a high temperature from an astrophysical standpoint.

Fireball radiation as a consequence of the big bang was predicted in 1948. Microwave measurements made by radio astronomers conform well to that predicted 3K pattern. If the 3K temperature is indeed a remnant of the big bang, then the whole history of the universe is out there for us to see, its boundaries fixed by ancient fireball radiation.

Distance measurements are based on redshift—the shifting of spectral lines toward the red end of the spectrum as a consequence of an object's recession. Might the shifting, though, be due to loss of energy by light as it passes through space, becoming "tired" in its encounter with matter? Hubble originally proposed the idea, and only recently support for the "tired light" hypothesis has been found, as a group of radio astronomers reported an apparent shifting of radio waves in passing through clusters of galaxies. If this proves to be so, then we may have to come back to the idea of an earlier static universe, a model that was abandoned when evidence for the expanding universe was found.

The momentum imparted in the big bang should decrease and expansion should slow down with time, just as objects hurled from the earth continue to decelerate. But according to another group of astronomers, the expansion is speeding up. Continual expansion at an accelerating rate would mean that the net force between galaxies is that of repulsion, bringing us back again to the "cosmological constant" of the earlier static model. Even the investigators themselves were surprised at their discovery, feeling that "something must be terribly wrong."

The observable universe is expanding. But when the force of the big bang is spent, will it all come collapsing back on itself billions of years from now, matter once again becoming radiation, the process beginning anew? Yes, say those who adhere to the oscillating universe theory, provided that there is sufficient material for gravity to exert its ultimate control. Recent studies, though, show that it can't happen: only one-tenth of the amount of material that is needed for this to be an oscillating universe can be found.

Present evidence favors the radiation-dominated origin of the universe in a big bang some 12 to 15 billion years ago. The observed expansion will continue forever. Nothing can stop it, for it has reached escape velocity. If we could watch the scene over tens of billions of years, we'd see the galaxies separating themselves farther and farther from each other, growing fainter, and finally disappearing beyond the horizon of visibility.

But if matter is being continuously created at the same rate that it is dispersing into the endless void—a theory, though, which observation does not strongly support—then the universe will remain as it is now. Forever.

# APPENDIX

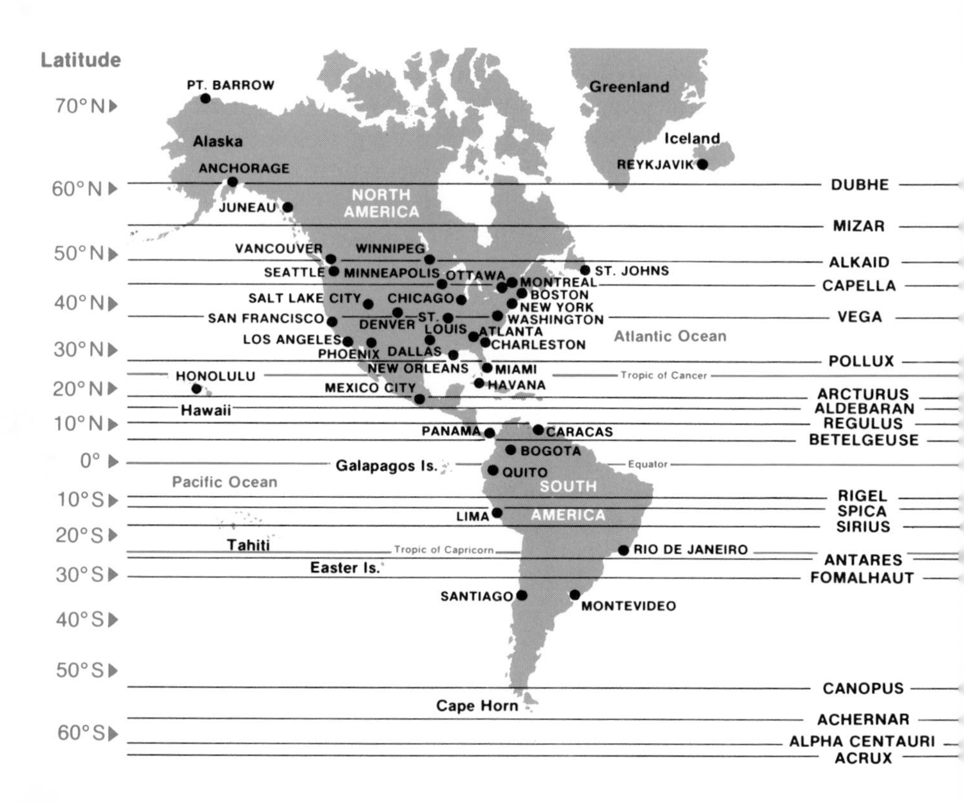


## ZENITH PASSAGE OF MAJOR STARS

The Earth spins. Stars travel in long arcs across the sky, moving always parallel to the equator.

An observer's zenith star passes directly over him once every twenty-four hours. Vega is the zenith star for many of the capital cities of the world, while Regulus is the zenith star for Panamanians, and Canopus for observers at the southern tip of South America.

The zenith passage of a star follows its line of *declination,* equivalent to latitude on Earth: The star is as many degrees north or south of the celestial equator as the observer is from the Earth's equator.

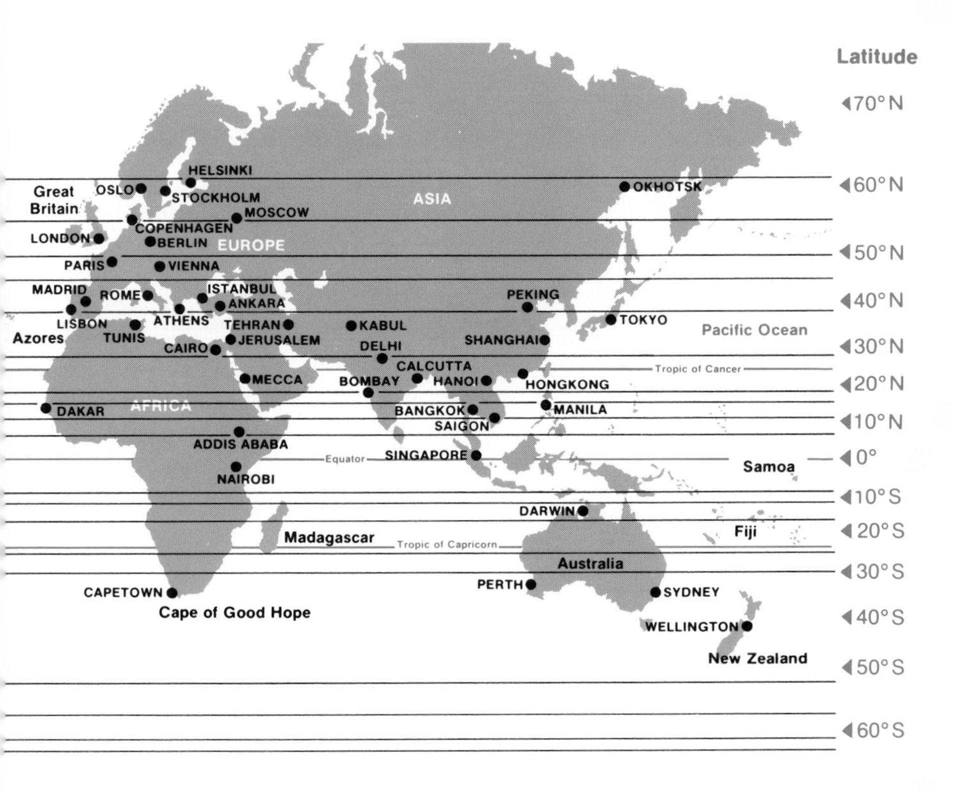


This zenith passage map can help you to determine which stars are at the limit of your horizon of visibility: Simply subtract your latitude from 90°; the difference will be the limit to which you can see into the opposite celestial hemisphere. Examples: Observers at 40° N, and 20° S:

90°
-40° N

50° S: You can see stars 50° south—not quite to Canopus.

90°
-20° S

70° N: You can see stars 70° north—slightly beyond the Big Dipper.

## THE BRIGHTEST STARS

| Name of Star | Distance (light-years) | Visual Magnitude | Absolute Magnitude |
|---|---|---|---|
| *Sun* | *0.000015 (8 minutes)* | *-26.73* | *4.84* |
| Sirius | 8.7 | -1.42 | +1.45 |
| Canopus | 98 | -0.72 | -3.1 |
| Alpha Centauri | 4.3 | -0.27 | +4.4 |
| Arcturus | 36 | -0.06 | -0.3 |
| Vega | 26.5 | +0.04 | +0.5 |
| Capella | 45 | 0.05 | -0.6 |
| Rigel | 900 | 0.14v | -7.1 |
| Procyon | 11.3 | 0.37 | +2.7 |
| Achernar | 118 | 0.51 | -2.3 |
| Beta Centauri | 490 | 0.63 | -5.2 |
| Altair | 16.5 | 0.77 | +2.2 |
| Betelgeuse | 520 | 0.41v | -5.6 |
| Aldebaran | 68 | 0.86v | -0.7 |
| Acrux | 370 | 0.9 | -4.0 |
| Spica | 220 | 0.91v | -3.5 |
| Antares | 520 | 0.92v | -5.1 |
| Pollux | 35 | 1.16 | +1.0 |
| Fomalhaut | 22.6 | 1.19 | +2.0 |
| Deneb | 1,600 | 1.26 | -7.1 |
| Beta Crucis | 490 | 1.28 | -4.6 |
| Regulus | 84 | 1.36 | -0.7 |

VISUAL MAGNITUDE: The brightness of the star as we see it in the sky.

ABSOLUTE MAGNITUDE: The brightness of the star as we'd see it at a distance of 32.6 light-years.

v = variable

## THE PLANETS

| Planet | Mean distance from Sun (millions of km) | Orbital Velocity (km/sec) | Equatorial Diameter (km) | Mass (Earth=1) | Density (Water=1) |
|---|---|---|---|---|---|
| Mercury | 57.9 | 47.8 | 4,880 | 0.055 | 5.4 |
| Venus | 108.2 | 35.0 | 12,104 | 0.815 | 5.2 |
| Earth | 149.6 | 29.8 | 12,756 | 1.0 | 5.5 |
| Mars | 227.9 | 24.2 | 6,787 | 0.108 | 3.9 |
| Jupiter | 778.3 | 13.1 | 142,800 | 317.9 | 1.3 |
| Saturn | 1,427.0 | 9.7 | 120,000 | 95.2 | 0.7 |
| Uranus | 2,869.6 | 6.8 | 51,800 | 14.6 | 1.2 |
| Neptune | 4,496.6 | 5.4 | 49,500 | 17.2 | 1.7 |
| Pluto | 5,900.0 | 4.7 | 6,000(?) | 0.1(?) | ? |

# INDEX OF CONSTELLATIONS

Based on the report by the Committee of the American Astronomical Society on Preferred Spellings and Pronunciations



*Shown on south circumpolar chart but not described in the text.

## INDEX OF STARS

Based on the report by the Committee of the American Astronomical Society on Preferred Spellings and Pronunciations



*Shown on star charts but not mentioned in text.

| | | |
|---|---|---|
| Deneb, *den' eb* | Cyg | 24-26, 67, 77, 144 |
| Denebola, *de-neb' o-la* | Leo | 53 |
| Diphda, *diff' da* | Cet | * |
| Dschubba, *jub' a* | Sco | * |
| Dubhe, *dub' ē* | UMa | 53, 142 |
| Elnath, *el' nath* | Tau | * |
| Eltanin, *el-tā' nin* | Dra | 69, |
| Fomalhaut, *fō' mal-ot* | PsA | 35, 42, 81, 83, 84, 142, 144 |
| Gacrux, *gā' kruks* | Cru | * |
| Gemma, *jem' a* (Alphecca) | CrB | 73 |
| Gienah, *jē' na* | Crv | * |
| Hamal, *ham' al* | Ari | 92 |
| Kaus Australis, *kos os-trā' lis* | Sgr | 81 |
| Kochab, *kō' kab* | UMi | * |
| Markab, *mar' kab* | Peg | 35, 84 |
| Megrez, *mē' grez* | UMa | 53 |
| Menkar, *men' kar* | Cet | * |
| Menkent, *men' kent* | Cen | * |
| Merak, *mē' rak* | UMa | 37, 53, 142 |
| Mintaka, *min' tak-a* | Ori | 39 |
| Mira, *mī' ra* | Cet | 81, 91, 128 |
| Mirach, *mī' rak* | And | 92 |
| Mirfak, *mir' fak* | Per | 92 |
| Mirzam, *mir' zam* | CaM | * |
| Mizar, *mī' zer* | UMa | 37, 53, 142 |
| Navi, *na' vi* | Cas | 88 |
| Phact, *fakt* | Col | * |
| Phecda, *fek' da* | UMa | 53 |
| Polaris, *po-lā' ris* ("North Star") | UMi | 7, 24-29, 35, 46, 59, 63, 67 78, 81, 83, 84, 88, 91, 92 |
| Pollux, *pol' uks* | Gem | 39, 45, 46, 142, 144 |
| Procyon, *prō' si-on* | CMi | 39, 49, 53, 109, 144 |
| Ras-Algethi, *ras' al-jē' thē* | Her | 70 |
| Rasalhague, *ras' al-hā' gwē* | Oph | 70 |
| Regulus, *reg' u-lus* | Leo | 42, 53, 55, 56, 83, 142, 144 |
| Rigel, *rī' jel* | Ori | 39, 45, 142, 144 |
| Ruchbach, *ruk' ba* | Cas | 88 |
| Sabik, *sā' bik* | Oph | * |
| Saiph, *sāf* | Ori | * |
| Scheat, *shē' at* | Peg | 35, 84 |
| Schedar, *shed' ar* | Cas | 88 |
| Shaula, *shaw' la* | Sco | * |
| Sheratan, *sher' a-tan* | Ari | * |
| Sirius, *sir' i-us* | CMa | 7, 11, 39, 46, 49, 50, 53 59, 73, 109, 127, 142, 144 |
| Spica, *spī' ka* | Vir | 53, 59, 64, 142, 144 |

## BIBLIOGRAPHY AND SUGGESTED READINGS


Abell, G.O. *Explorations of the Universe.* New York: Holt, Rinehart & Winston, 1975.

Allen, R.H. *Star Names.* New York: Dover, 1963.

Alter, D., Cleminshaw, C.H., and Phillips, J. *Pictorial Astronomy.* New York: Crowell, 1974.

Bok, B.J., and Bok, P.F. *The Milky Way.* Cambridge, Mass.: Harvard University Press, 1973.

Bova, B. *The New Astronomies.* New York: New American Library, 1972.

Brown, P.L. *Astronomy.* New York: Macmillan, 1972.

Calder, N. *Violent Universe.* New York: Viking Press, 1970.

Charon, J. *Cosmology.* New York: McGraw-Hill, 1970.

Clark, E.E. *Indian Legends of the Pacific Northwest.* Berkeley: University of California Press, 1953.

Colum, P. *Myths of the World.* New York: Grosset & Dunlap, 1930.

Cumont, F. *Astrology and Religion among the Greeks and Romans.* New York: Dover, 1960.

Gamov, G. *Thirty Years that Shook Physics.* Garden City: Doubleday, 1966.

Hamilton, E. *Mythology.* New York: Mentor Books, 1940.

Hawkins, G.S. *Stonehenge Decoded.* New York: Doubleday, 1965.

Hawkins, G.S. *Beyond Stonehenge.* New York: Harper & Row, 1973.

Hodge, P.W. *Concepts of the Universe.* New York: McGraw-Hill, 1969.

Hoyle, F. *Frontiers of Astronomy.* New York: Harper & Row, 1955.

Hoyle, F. *Astronomy.* New York: Doubleday, 1962.

Hubble, E. *The Realm of the Nebulae.* New York: Dover, 1958.

Kaufmann, W.J. *Relativity and Cosmology.* New York: Harper & Row, 1973.

Kilmister, C.W. *The Nature of the Universe.* New York: Dutton, 1971.

Komaroff, K. *Sky Gods, the Sun and Moon in Art and Myth.* New York: Universe Books, 1974.

Kuhn, T. *The Copernican Revolution.* New York: Vintage Books, 1959.

Levitt, I.M., and Marshall, R.K. *Star Maps for Beginners.* New York: Simon & Schuster, 1964.

Lovell, B. *Exploration of Outer Space.* New York: Harper & Row, 1962.
Martin, M.E., and Menzel, D.H. *The Friendly Skies.* New York: Dover, 1964.
Menzel, D.H. *A Field Guide to the Stars and Planets.* Boston: Houghton Mifflin, 1964.
Nangle, J. *Stars of the Southern Heavens.* Sydney: Angus & Robertson, 1962.
Neugebauer, O. *The Exact Sciences in Antiquity.* New York: Dover, 1969.
Norton, W.W. *Sky Atlas.* Cambridge, Mass.: Sky Publishing, 1971.
Olcott, W.T. *Olcott's Field Book of the Skies.* New York: Putnam, 1954.
Roach, F.E., and Gordon, J.L. *The Light of the Night Sky.* Dordrecht, Holland: Reidel, 1973.
Roberts, A., and Mountford, C.P. *The Dawn of Time.* Adelaide: Rigby, 1969.
Sagan, C. *The Cosmic Connection.* Garden City, N.Y.: Anchor Press Doubleday, 1973.
Santillana, G. de, and Dechend, H. von. *Hamlet's Mill.* Boston: Gambit, 1969.
Shapley, H. *Galaxies.* Cambridge, Mass.: Harvard University Press, 1972.
Shklovskii, I.S., and Sagan, C. *Intelligent Life in the Universe.* San Francisco: Holden-Day, 1966.
Silva, J.L., and Lochak, G. *Quanta.* New York: McGraw-Hill, 1969.
Singh, J. *Great Ideas and Theories of Modern Cosmology.* New York: Dover, 1961.
Thiele, R. *And There was Light.* New York: Alfred A. Knopf, 1957.
Thompson, C.J.S. *The Mystery and Romance of Astrology.* New York: Causeway Books, 1973.
Warner, Lionel. *Astronomy for the Southern Hemisphere.* Wellington: A.H. & A.W. Reed, 1975.
Whitrow, G.J. *The Structure and Evolution of the Universe.* New York: Harper, 1959.

Journals and Periodicals:

*Astronomy,* AstroMedia Corp., Milwaukee, Wisconsin.
*The Griffith Observer,* Griffith Observatory, Los Angeles, California.
*Mercury,* Astronomical Society of the Pacific, San Francisco, California.
*Scientific American,* Scientific American, New York.
*Sky and Telescope,* Sky Publishing Corporation, Harvard College Observatory, Cambridge, Massachusetts.

# GENERAL INDEX